BIOCHEMISTRY For MEDICAL STUDENTS

Dr. N Haridas PhD (Medical Biochemistry)
Professor and Head, Department of Biochemistry
Pramukhswami Medical College
and Shrikrishna Hospital
Karamsad 388 325, Gujarat

New Central Book Agency (P) Ltd
LONDON
HYDERABAD ERNAKULAM BHUBANESWAR
NEW DELHI KOLKATA PUNE GUWAHATI

NCBA

REGD OFFICE

8/1 Chintamoni Das Lane, Kolkata 700 009, India
email: ncbapvtltd@eth.net

OVERSEAS

NCBA (UK) Ltd, 149 Park Avenue North, Northampton, NN3 2HY, UK
email: ncbauk@yahoo.com

EXPORT

NCBA Exports Pvt. Ltd, 212 Shahpur Jat, New Delhi 110 049
email: ncbaexports@ncbapvtltd.com

BRANCHES

2/27 Ansari Road, 1st Floor
Daryaganj, **New Delhi** 110 002
email: ncbadel@ncbapvtltd.com

House No. 3-1-315, 1st Floor
Nimboliadda, Kachiguda, **Hyderabad** 500 027
email: ncbahydb@ncbapvtltd.com

Shop Nos. 3 & 4, Vinayak Towers
681/B Budhwar Peth, Appa Balwant Chowk
Pune 411 002
email: ncbapune@ncbapvtltd.com

GSS Shopping Complex, 1st Floor
Opposite College Post Office, Convent Road
Ernakulam 682 035
0484-4021054/53

Shop No. 15, Bharati Towers, Ground Floor
Forest Park
Bhubaneswar 751 001
email: ncbabub@ncbapvtltd.com

Radhamayee Complex (Second Floor)
Opp. SBI, Panbazar, Jaswanta Road
Guwahati 781 001
email: ncbaguwahati@ncbapvtltd.com

Biochemistry for Medical Students • Dr. N Haridas

First Published: July 2015
Reprinted: March 2016

PUBLISHER AND TYPESETTER
New Central Book Agency (P) Ltd
8/1 Chintamoni Das Lane, Kolkata 700 009

PRINTER
New Central Book Agency (P) Ltd
Web-Offset Division, Dhulagarh, Sankrail, Howrah

TECHNICAL EDITORS
Dr Mita Sen MBBS
Dipan Roy

PROJECT SUPERVISOR
Prabir Kumar Ghosh

ISBN 978 81 7381 963 6

Price ₹ 185.00

Dedicated to
my Late Parents
and
all the elders of my family

BIOCHEMISTRY
For
MEDICAL STUDENTS

CONTENTS

CONTENTS

PREFACE

After having had the opportunity to interact with umpteen number of undergraduate and postgraduate students over the past so many years, the thought of putting my experiences on paper dawned on me and this work is the result.

This treatise is expected to satisfy the needs of the MBBS students as far as the learning of Biochemistry is concerned. It may be argued that there are several books on Biochemistry catering to the need of medical students, then why another book? Most of such books are voluminous. A medical student who gets only one year time to study the subjects of Anatomy, Physiology and Biochemistry—along with a part of Community Medicine—could be hard pressed to master all these subjects in such a short span of time. Hence the need of the hour is to have a concise book mainly focussing on the applied aspects of the topics along with a brief description of the basic concepts, required to understand and interpret the information—especially in health and disease. This book is an humble attempt in that direction.

The language employed is simple, lucid and easy to follow as many of the students joining the MBBS course could have studied through vernacular medium. This book can come in handy to other paramedical courses like Physiotherapy, Medical Technology and Pharmacy as well.

23rd January 2015
Kolkata

Dr N Haridas

ACKNOWLEDGEMENTS

I hereby express my sincere thanks to New Central Book Agency (P) Ltd, Kolkata, for publishing this treatise "Biochemistry for Medical Students".

I am grateful to the Charutar Arogya Mandal and its Dean for giving permission to author this book and for the encouragement extended to me.

Further, I have to thank my beloved wife for giving me good support in completing this work.

The students numbering in thousands have been a constant inspiration to me in sharpening my teaching skills, helping me to become a passionate teacher, and, therefore, they deserve sincere thanks on my behalf. Their love and respect enabled me to conduct the teaching-learning sessions optimally over the last 37 years. No wonder that I have enjoyed my profession immensely and all the credit for this goes to my students only.

I express my sincere thanks to Mr Amitabha Sen, Director, New Central Book Agency (P) Ltd, for his personal interest and efforts at all stages of production. My special thanks are due to Mr Prabir Ghosh, project supervisor, for his active role in publishing the book in time.

chapter 1

Carbohydrates

The carbohydrates are defined as polyhydroxy aldehydes or ketones. They are classified into monosaccharides, disaccharides and polysaccharides.

Monosaccharides

The monosaccharides can be trioses, tetroses, pentoses, hexoses and heptoses depending on the number of carbons. An important tetrose is erythrose. The important pentoses are ribose, ribulose and xylulose. The important hexoses are glucose, fructose, galactose and mannose. One example of heptose is heptulose, a 7-carbon sugar with a keto group.

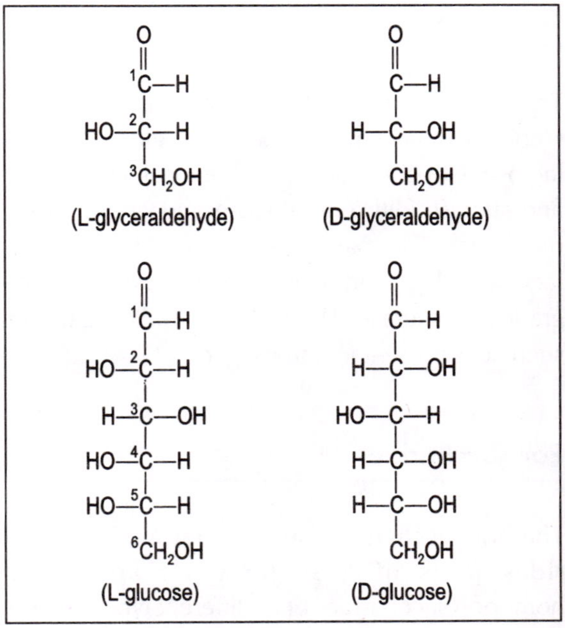

Fig. 1.1: Isomerism in monosaccharides

Out of all the monosaccharides, glucose is the most important as it is the major energy fuel in the body, its blood level being maintained at all costs, because tissues like brain and erythrocytes can utilise only glucose as the energy source. Hence, the blood glucose level should be within 70-110 mg/dl in the blood. A failure to maintain this could have far reaching consequences like hyperglycaemia (diabetes mellitus) or hypoglycaemia.

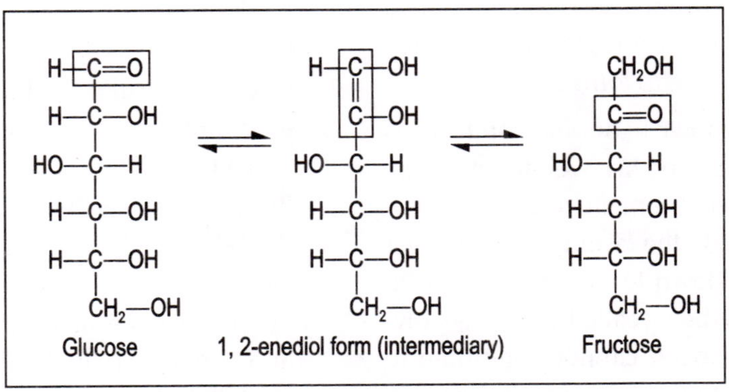

Fig. 1.2: Interconversion of glucose and fructose

Disaccharides

The disaccharides contain two monosaccharides. Hence, we have maltose with two glucose units, lactose with one glucose and one galactose units and sucrase with one glucose and one fructose molecules. The maltose is degraded by maltase, lactose is degraded by lactase and sucrose is degraded by sucrase. The lactose is milk sugar and sucrose is present in cane sugar.

Polysaccharides

The polysaccharides have several monosaccharides units of a single type, known as homopolysaccharides or of different types, called as heteropolysaccharides, also called as mucopolysaccharides or heteroglycans.

Homopolysaccharides

Starch in plants and glycogen in animals are examples of homopolysaccharides. Both are made up of several glucose units in alpha-1, 4 glucosidic linkages and alpha-1, 6 glucosidic bonds at the branch points. The starch contains amylase and amylopectins the latter being more branched than the former. The main ingredient of an Indian diet is carbohydrate, that too mainly starch. The starch is hydrolysed by amylase present in saliva (negligible) and pancreatic juice to maltose which is acted upon by maltase to form glucose.

The glycogen is known as animal starch and is stored in muscles and liver. In a well-fed state, the glycogen is synthesised from glucose by the process of glycogenesis. In the fasting state, the liver glycogen is broken down to provide glucose to the blood, the process being called as glycogenolysis. The muscle glycogen cannot contribute to blood glucose, due to the absence of glucose-6 phosphatase enzyme in the muscles. The muscle glycogen can provide glucose-6 phosphate for muscle activities.

The cellulose is also made up of glucose units, but by beta 1,4 linkages and has no branch points. Humans do not have the enzyme cellobiase to break the beta linkages.

The inulin is a homopolysaccharide containing fructose units and is present in onions and tubers. Inulin finds use in measuring glomerular filtration rate and renal clearance.

Another homopolymer, synthesised by bacteria are dextrans with glucose residues in a very branched structure. It is used for the treatment of hypovolemic shock as it can remain in the vessels for considerable time.

Heteropolysaccharides/ Mucopolysaccharides

The heteropolysaccharides of maximum importance are known as mucopolysaccharides. They are glycosaminoglycans and are made up of uronic acid and amino sugars. The mucopolysaccharides can form viscous solutions. They can form mucoproteins by joining proteins. The examples of mucopolysaccharides are given below:

1. Hyaluronic acid: It is composed of N-acetyl-glucosamine and 1, 4-glucuronic acid, repeated several times. It is present in the vitreous humour, synovial fluid, connective tissue, tendons and cartilages. The hyaluronic acid has lubricant properties.
2. Chondroitin sulphate: It consists of glucuronic acid and N-acetyl galactosamine sulphate, repeated several times. This can be seen in the ground substance of bones, cartilages, tendons, skin, etc.
3. Heparin is an anticoagulant which is used both in vitro (for collecting blood to prepare plasma) and also in vivo (for treatment purposes). It is made up of repeated units of sulphated glucosamine and iduronic acid.

4. The dermatan sulphate is made up of N-acetyl galactosamine and iduronic acid. This mucopolysaccharide is seen in the skin and arteries.

5. The keratan sulphate consists of repeated residues of galactose and N-acetyl glucosamine and these are components of tendons and cartilages. These glycosamino glycans do not have uronic acids.

chapter 2

PROTEINS AND AMINO ACIDS

The proteins are polymers of amino acids. There are 20 types of amino acids in the proteins. The function of the protein depends on the content and the sequence of amino acids. The amino acids, in addition to being the building blocks of proteins, have several other roles like serving as precursors of many compounds like dopamine, epinephrine, norepinephrine, thyroxine, melanine, GABA, serotonin, heme, creatinine and so on. Some of the amino acids also serve as neurotransmitters.

AMINO ACIDS

The amino acids can be classified in different ways. One type of classification is as follows:

1. **Neutral amino acids**: These possess a negatively charged carboxyl group and a positively charged amino group. The example of this group are glycine, alanine, valine, leucine and isoleucine. Out of these, glycine is the smallest amino acid and is required for heme and glutathione synthesis. Further, every third amino acid in collagen is glycine.
2. **Amino acids with hydroxyl group**: Serine and threonine. Some of the enzymes can be phosphorylated at their hydroxyl groups so as to make them active or inactive, depending on the enzymes. For example, glycogen phosphorylase, the rate limiting enzyme of glycogenolysis becomes active when phosphorylated, whereas the glycogen synthase, the rate limiting enzyme of glycogenesis is made inactive by phosphorylation of the hydroxyl group of serine amino acid.
3. **Amino acids with –SH groups**: The amino acids with sulphur are cysteine and methionine. The cysteine is involved in the formation of interchain and intrachain disulfide bridges in polypeptide chains. One example is the disulfide bridges of insulin and immunoglobulins. The methionine is the first amino acid to be incorporated into any protein and it is involved as a methyl group donor, the active methionine being S-adensyl methionine.
4. **Acidic amino acids**: The examples of acidic amino acids are aspartic acid and glutamic acid. They contain two carboxylic acid groups, and hence, have a net negative charge. A protein containing more of acidic amino acids will be acidic in nature. The sickle cell haemoglobin results when the 6th amino acid on the beta chain of haemoglobin is changed from acidic glutamic acid to neutral valine. The glutamic acid is decarboxylated to gamma amino butyric acid, a neurotransmitter.
5. **Amide group containing amino acids**: Several proteins have two amide groups containing amino acids, asparagine and glutamine which are rather rare in proteins. The increase in the number of glutamine residue in a protein, known as Huntington protein, causes a hereditary disease known as Huntington chorea, characterised by involuntary jerky movements. The glutamine is converted to glutamic acid and ammonia by glutaminase, present in kidneys which assumes significance in the regulation of acid-base balance. The glutamine is synthesised by a different enzyme, the glutamine synthase in presence of ATP. The asparaginase which cleaves aspar-

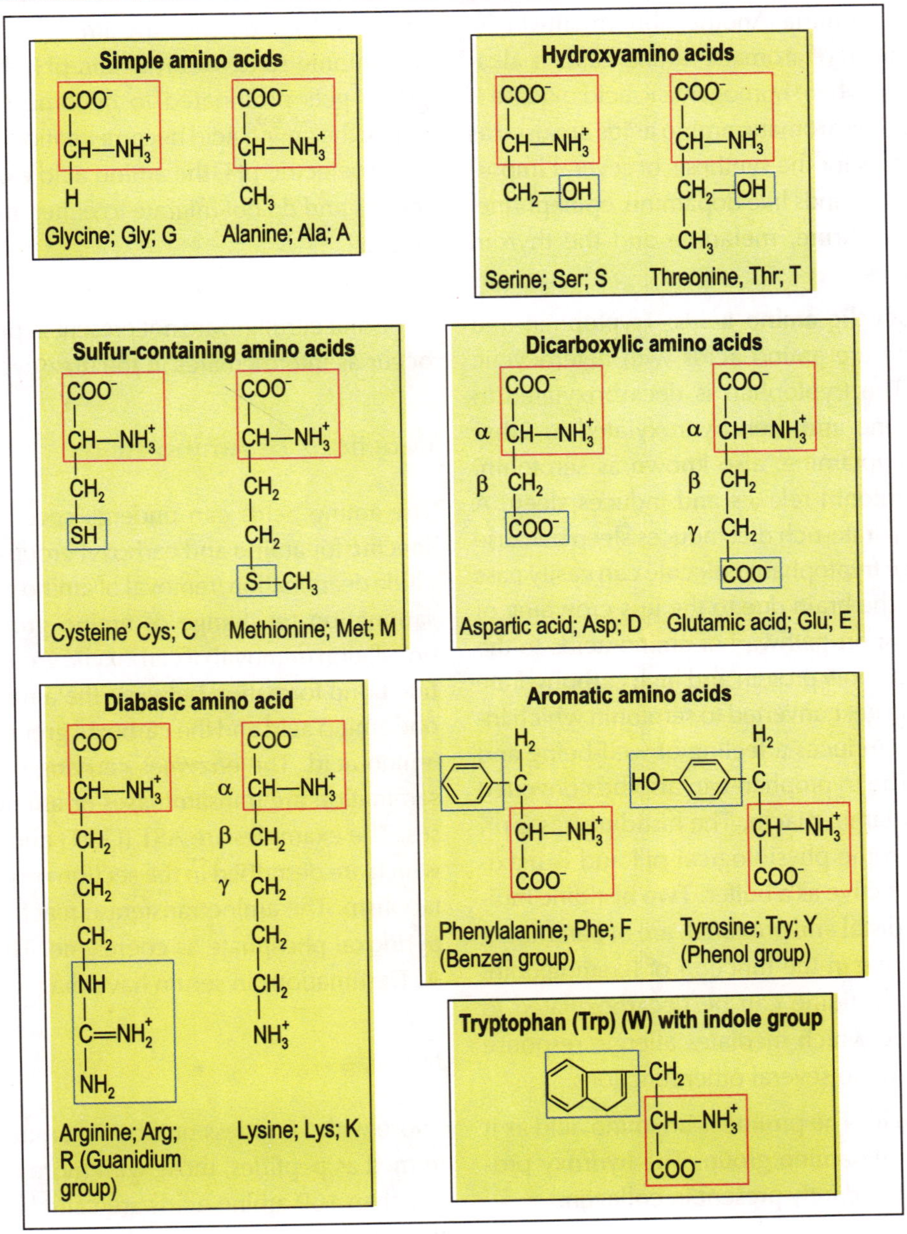

Fig. 2.1: Structure of amino acids

agine is used as a anticancer agent, as the cancer cells are starved off of this amino acid.

6. **Basic amino acids**: The basic amino acids are lysine and arginine which have extra amino groups. The histones, which interact with DNA, are basic proteins that are rich in lysine and arginine. Arginine is required for the synthesis of creatine. The structural protein collagen is rich in hydroxy lysine.

7. **Aromatic amino acids**: The aromatic group containing amino acids are phenyl alanine and tyrosine. The hydroxylation of phenyl alanine by phenyl alanine hydroxylase produces tyrosine. The absence of this enzyme causes

phenyl ketonuria. Another abnormality in the metabolism of aromatic amino acids is alkaptonuria, where homogentisic acid oxidase is absent. The aromatic amino acids are used as precursors for the synthesis of several important compounds like dopamine, epinephrine, norepinephrine, melanine and the thyroid hormones.

8. **Heterocyclic amino acids**: Tryptophan and histidine are amino acids with heterocyclic rings. The tryptophan is decarboxylated to tryptamine and then hydroxylated to 5-hydroxy tryptamine, also known as serotonin. The serotonin relaxes and induces sleep. A carbohydrate-rich diet induces sleep because the large tryptophan molecule can easily pass over to the brain due to the less crowding of the transport pathway of amino acids to the brain, in a low protein and high carbohydrate diet, and get converted to serotonin which invariably induces a feeling of well-being and sleep. The tryptophan can also be converted to the vitamin niacin. The histidine has 2 pK values in the physiological pH and is maximum effective as a buffer. Two histidine molecules, distal and proximal, are of paramount importance in the function of haemoglobin. Further, histidine can be decarboxylated to histamine which mediates allergic response in addition to several other functions.
9. **Imino acid**: The proline is an imino acid as it has no true amino group. The hydroxy proline is abundantly present in collagen.

The amino acids can remain in 3 forms: zwitterions, in which, the amino acid remains neutral as both the amino group and carboxylic acid group are ionised, having positive and negative charges, thereby forming electrically neutral zwitterions. This happens at the isoelectric pH (pK) of the amino acid. The pH of the medium decides the electrical charges on the amino acid. At alkaline pH, the carboxyl group of the amino acid gets ionised to COO^- (with the amino group remaining as NH_2) and hence the amino acid will be in the anionic state. At an acidic pH, the carboxyl group gets protonated to become COOH and hence the amino acid becomes cationic with NH_3^+. At isoelectric pH, the amino acid exists as zwitterions and do not migrate to either anode or cathode.

Amino acids like argino succinic acid, ornithine and citrulline are not present in proteins, but, occur as intermediates in the urea cycle.

Reactions of Amino Acids

The amino acids can undergo several reactions specific for amino and carboxyl groups. These include deamination (removal of amino group), transamination (exchange of amino group), decarboxylation (removal of carboxylic group) and peptide bond formation between the amino group of one amino acid and the carboxyl group of another amino acid. The enzymes carrying out the transamination are transaminases or aminotransferases. The examples are AST (GOT) and ALT (GPT) which are described in the section on protein metabolism. The aminotransferase reactions require pyridoxal phosphate as coenzyme. The AST and ALT estimations in serum have diagnostic value.

Peptides

Those with 10 or less number of amino acids are termed as peptides, those with more than 10 but less than 100 amino acids are called as polypeptides and those with more than 100 amino acids are known as proteins, although this is an arbitrary way of classification.

The examples of peptides consist of the thyrotropin releasing hormone (TRH-a dipeptide), glutathione (a tripeptide, gamma glutamyl cysteinyl glycine), vasopressin, oxytocin (both with 9 amino acids), angiotensin 1 and 2 (with 8 amino acids) and endorphins and enkephalins, which have analgesic properties.

Polypeptides

The insulin and glucagon are polypeptide hormones having 51 and 29 amino acids respectively. They regulate the metabolism of carbohydrate, lipid and protein. The levels of these hormones are crucial for the maintenance of blood glucose and the whole of the metabolic reactions.

PROTEINS

Proteins are one of the most dynamic molecules performing a variety of functions, and so they are better classified based on their functions.

1. **Catalytic proteins**: The enzymes are biocatalysts and bring about the biochemical reactions in an organism. All the enzymes (except a very few RNA containing ribozymes) are proteins which are highly specific in their action. The enzymes carry out a variety of reactions like oxidoreductions, transfer of certain groups from one substrate to another, hydrolysing macromolecules into small molecules by the addition of water, simply splitting a compound into two without the involvement of water, interconversion of isomeric compounds and joining two substrates, the enzymes carrying out these processes being named respectively as oxidoreductase, transferase, hydrolase, lyase, isomerase and ligase. The specificity, the property that an enzyme can act only on one substrate or a group of similar substrates, is another hallmark of an enzyme. The specificity is so stringent that it extends to the requirement of specific isomers of the substrate and co-enzymes also.
2. **Transport proteins**: They transport various substances in the blood/plasma. The example of a transport protein is haemoglobin that transports oxygen from lungs to the tissues and carbon dioxide from the tissues to the lungs. Another example is that of plasma albumin which transports several substances like non-esterified fatty acids, calcium, copper, bilirubin, drugs and steroids. All the lipoproteins are also transport proteins as they are involved in lipid transport. Transferrin, retinol binding protein and thyroxine binding globulins transport iron, retinol and thyroxine respectively.
3. **Structural proteins** form the matrix of the tissues and keep the cells of the tissues together. The examples are keratin of hair, collagen of bones, tendons, cartilages, walls of the blood vessels, skin and lens of the eyes, the type of the collagen of different tissues having slightly altered structures. The elastin is another structural protein.
4. **Defense proteins** are immunoglobulins belonging to 5 major classes: IgG, IgA, IgM, IgD and IgE. In general, these proteins confer resistance to infections. They are synthesised by B-lymphocytes.
5. **Hormonal proteins**: The examples of this type of proteins are growth hormone, insulin, glucagon prolactin etc. Certain hormones are proteins, but all the hormones are not proteins.
6. **Metabolic regulatory proteins**: Calmodulin can bind with calcium and can regulate metabolic processes like glycogen metabolism.
7. **Gene regulatory proteins**: The histone and nonhistone proteins that bind to DNA in the major and minor grooves to form nucleoproteins regulate the expression of genes, like DNA replication and transcription.
8. The muscles have contractile proteins, like actin and myosin.

Structure of Proteins

The proteins have primary, secondary, tertiary and quaternary levels of structures.

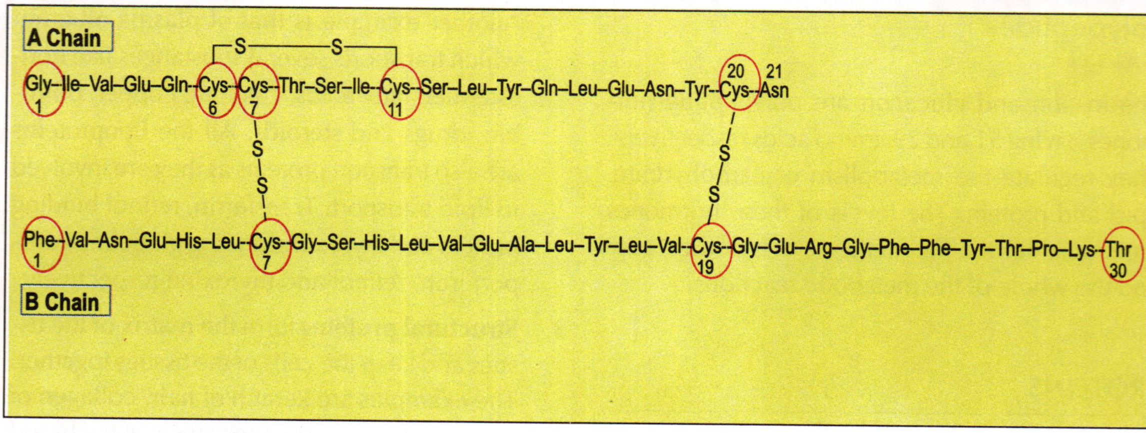

Fig. 2.2 Amino acid sequence of human insulin

Primary Structure of Proteins

The primary structure of proteins is a linear sequence of amino acids and is made up of peptide bonds. The sequence of amino acid in a protein is dictated by the sequence of nucleotides in the corresponding gene. The primary structure of the first protein to be determined was that of insulin and was done by Sanger who got Nobel prize for the same. He used a reagent by name fluoro dinitro benzene (FDNB) to react it with the N-terminal amino acid and subsequently recognising it by chromatography after removing that amino acid. Now the original second amino acid becomes the N-terminal amino acid and the reaction with FDNB and the recognition process is repeated. Like this by several steps Sanger was able to sequence the whole of insulin. It was elucidated that the insulin has two polypeptide chains, the A and B chains connected by disulphide bonds. The A chain has 21 amino acids and the B chain has 30 amino acids. One inter-chain disulphide bond connects the 7th cysteine of A chain with the 7th cysteine of B chain. Another interchain disulphide bond is between the 20th cysteine of A chain and the 19th cysteine of the B chain. The intra- chain disulphide is between the 6th cysteine and the 11th cysteine of the A chain. The active insulin is formed from the pro-insulin by the removal of the connecting peptide. The insulin molecules from various animals like ox, sheep and horse are different in the sequence of amino acids at 8th-10th amino acids of the A chain. The porcine insulin is almost similar to human insulin except the 30th amino acid in B chain which is alanine in porcine insulin while threonine is present in human insulin. In the human insulin, the N-terminal amino acid is glycine and the C-terminal amino acid is asparagine in the A chain. In the B chain the N-terminal amino acid is phenyl alanine and the C-terminal amino acid is threonine.

Secondary Structure of Proteins

The primary structure is a linear one and the secondary structure is a helical one. There are three types of secondary structures. One is known as alpha helix which is formed by hydrogen bonding between every 4th amino acid. The hydrogen bond is formed by peptidyl nitrogen and carbonyl oxygen. Because of this the polypeptide chain is twisted to form the helical structure. The alpha helix is further stabilised by salt bonds formed by the extra amino groups of basic amino acids and the extra carboxyl groups of acidic amino acids. The van der Waals forces, the force of attraction and repulsion between atoms also help to stabilise the alpha helix. Further, the hydrophobic interactions between neutral amino acids which try to come closer avoiding water molecules have also got a major say in the formation of higher order structures of proteins. The pitch of one turn of helix is 0.54nm and there are 3.6 amino acids in one turn of the helix.

The second type of secondary structure is beta-pleated sheet which is a stretched structure. Here the hydrogen bonds are formed by amino acids which are apart from each other, like, say, every 10th amino acid. The other types of bonds/interactions like salt bridges, hydrophobic interactions and van der Waals forces, etc. are also important here.

Those areas of protein structure which have no organised level of structure are known as random coil. It does not mean that random coil is less important.

Tertiary Structure

It is the interaction between different domains or areas of the polypeptide chain. In this case interactions may be between amino acids located quite apart. The tertiary structure is made up of a single polypeptide chain which is coiled to take a three-dimensional structure. The various type of interactions involved in stabilising the secondary structure is important in tertiary structure also. Albumin and myoglobin have a tertiary level structure.

Quaternary Structure

The quaternary structure requires more than one polypeptide chain or subunit or monomers which are combined together to form a dimer or a tetramer. The monomers are not active and all the subunits are required for the protein to be active.

The examples consist of immunoglobulins, with 2 light and 2 heavy chains, haemoglobin with 2 alpha and 2 beta chains and LDH4 with 3M chains and 1H chain. If the quaternary structure is degraded, the monomers will not be active.

When the proteins are heated or treated with alcohol or other solvents, the protein loses all higher order structures like quaternary, tertiary and secondary level of structures with only primary structure remaining undisturbed.

The supra-quaternary level is obtained in multienzyme systems, like alpha keto glutarate dehydrogenase, pyruvate dehydrogenase and fatty acid synthase complex. In such systems the enzymes of a particular metabolic pathway are grouped together, so as to avoid interfering pathways and improve the efficiency of the pathway.

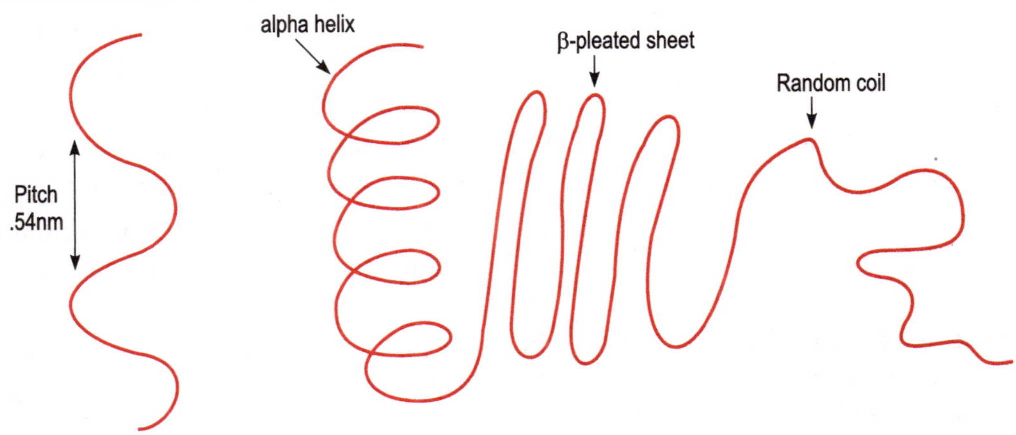

Alpha-helix structure

Secondary structue

chapter 3

Lipids

The lipids or fats are in general soluble in fat solvents like alcohol, ether, acetone, etc. but insoluble in water. They are heterogeneous class of compounds and consist of simple lipids, compound lipids and derived lipids.

SIMPLE LIPIDS OR TRIACYL GLYCEROL

The simple lipids are esters of fatty acids and alcohols like glycerol and other higher alcohols. The examples are neutral fat or triacyl glycerol (TAG) or triglyceride (TG) which have a glycerol esterified with 3 fatty acids which may be the same (simple TAG) or different (mixed TAG). The oils are liquid and the fats are solid at room temperature.

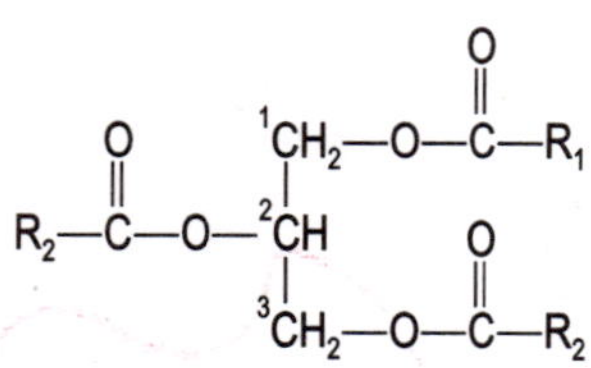

FATTY ACIDS

The fatty acids with double bond/s (—CH=CH—) are known as unsaturated fatty acids and the ones without double bond are known as saturated fatty acids. The saturated fatty acids are more present in non-vegetarian foods and the unsaturated ones are more in vegetable oils. Clinically speaking, the unsaturated fatty acids are good and the saturated ones, especially the long chain ones are bad for health, if taken in excess. The saturated fatty acids include palmitic and stearic acid.

Unsaturated Fatty Acids

The unsaturated fatty acids exhibit cis-trans isomerism. All the unsaturated fatty acids seen in the nature are of cis variety. The examples include oleic acids, linoleic acid, linolenic acid and arachidonic acid—the last three are essential fatty acids. The unsaturated fatty acids can be either mono unsaturated (MUFA) or poly unsaturated (PUFA), which can be either omega-3 or omega-6.

Trans-fatty Acids

The trans-fatty acids, produced during the preparation of fried and baked foods, are injurious to health as they increase the harmful low density lipoproteins and decrease the beneficial high density lipoproteins. Although trans-fatty acids are formed in the body during the beta oxidation of fatty acids these do not accumulate as they are consumed in the next reaction step in the pathway.

Another method of classification of fatty acids, based on their carbon chain length is short chain (2-6 carbons), medium chain (8-14 carbons), long chain (16-24 carbons) and very long chain fatty acids (more than 24 carbons).

Yet another way of classifying fatty acids is odd and even carbon number fatty acids.

Solid fats mainly contain long chain saturated fatty acids while oils have more of unsaturated fatty acids.

Non-vegetarian Fats

The non-vegetarian fats, like red meat are rich in long chain saturated fatty acids like stearic and

Polyunsaturated fatty acids (PFUA)
Linoleic (C_{18})Δ9, 12 (two double bonds) (ω6 family) CH_3—$(CH_2)_4$—CH=CH—CH_2—CH=CH—$(CH_2)_7$—COOH 18 ω6 12 9 1
Linoleic (C_{18})Δ9, 12, 15 (three double bonds) (ω3 family) CH_3—CH_2—CH=CH—CH_2—CH=CH—CH_2—CH=CH—$(CH_2)_7$—COOH 18 ω3 15 12 9 1
Arachidonic (C_{20})Δ5, 8, 11, 14 (four double bonds) (ω6 family) CH_3—$(CH_2)_4$—CH=CH—CH_2—CH—CH—CH_2—CH=CH—CH_2—CH=CH—$(CH_2)_3$—COOH 18 ω6 14 11 8 5 1

palmitic acids which are bad for health when consumed in excess as they increase the serum low density lipoprotein cholesterol values. Milk and dairy products are rich in short and medium saturated fatty acids which do not alter the cholesterol concentrations and hence is harmless. The fish oils are an exception as they are rich in polyunsaturated fatty acids especially docosa hexaenoic acid which lowers the LDL- cholesterol, while increasing the good high density lipoprotein cholesterol and hence have beneficial effect.

Vegetarian Oils

The vegetable oils have more of poly unsaturated fatty acids like linoleic (18 carbons with 2 double bonds), linolenic (18 carbons with 3 double bonds) and arachidonic (20 carbons with 4 double bonds) acids which are good for health as they reduce the serum cholesterol levels thereby preventing Coronary artery disease. The linoleic acid and arachidonic acid are omega-6 and linolenic acid is omega-3 fatty acid. The arachidonic, the precursor of prostaglandins, can be synthesised from linoleic acid. These are known as essential fatty acids as they cannot be formed in the body except arachidonic acid. A healthy proportion of PUFA and omega-3 fatty acids are present in olive oil and soyabean oil. The coconut oil is an exception which in spite of being a vegetable oil has more of saturated fatty acids, but these are of short and medium chain fatty acids. The digestion and absorption and the metabolic fate of short chain, medium chain, long chain and very long chain fatty acid containing TAGs are different.

The medium and short chain fatty acids containing TAGs do not need bile salts (emulsifying agent) and pancreatic lipase (requiring only gastric and lingual lipases) for digestion and absorption. They do not need chylomicron formation and their products of digestion are water soluble and hence, are directly absorbed into the blood and not via the lymphatics, as is the case with those TAGs with long and very long chain fatty acids. Further the SCFA and MCFA do not affect serum cholesterol levels and are not harmful. The SCFA and MCFA can be given to patients with type-1 hyperlipoproteinaemia due to the deficiency of lipoprotein lipase required to remove chylomicrons from the plasma.

Phospholipids

The phospholipids contain 1 glycerol, 2 fatty acids and a nitrogenous base.

Lecithin and Cephalin

The phosphatidyl choline is known as lecithin. One PUFA is attached to the 2nd carbon of glycerol. The enzyme, phospholipase-A_2 acts on the

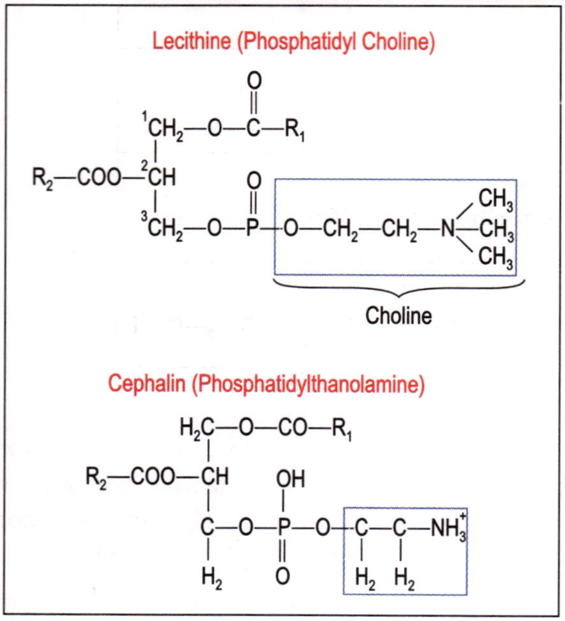

lecithin to remove the fatty acid at carbon-2 of glycerol and forms lysolecithin, a haemolytic agent, present in snake venom. The dipalmityl lecithin functions as a lung surfactant preventing the collapse alveoli. In premature infants, the deficiency may cause respiratory distress syndrome. The phosphatidyl ethanolamine is called as cephalin. Both lecithin and cephalin are present in biomembranes and in nervous tissue. These two have both polar and non-polar groups.

Phosphatidyl Inositol

These are also constituents of biomembranes. There is an inositol in place of the nitrogenous base. The phosphatidyl inositol diphosphate is a mediator hormone action.

Plasmalogens

The other phospholipids include plasmalogens containing an ether linkage. There is a higher alcohol and nitrogen base also. These are constituents of brain.

Cardiolipins

The cardiolipins seen in myocardium have glycerol and two moles of phosphatidic acid.

Sphingolipids

The sphingolipids have an amino alcohol, sphingosine and fatty acid. There are glycosphingolipids and sulphatides. The sphingolipids containing phosphate but no sugar are known as sphingomyelins, abundantly present in nervous tissue. The lecithin/sphingomyelin ratio in amniotic fluid is very important to determine the lung maturity in newborns.

Glycosphingolipids

Again present in nervous tissue, they contain sphingosine, fatty acid (which forms ceramide) and glucose or galactose to form glucocerebrosides and galactocerebrosides respectively.

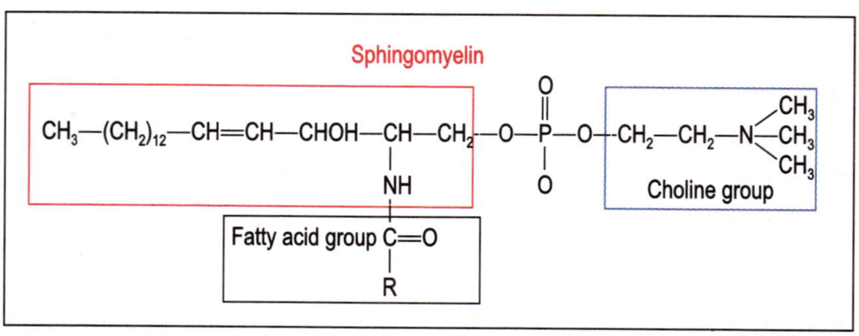

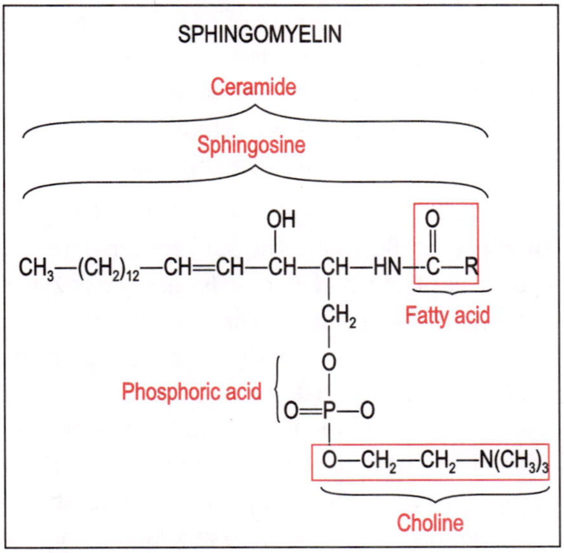

Gangliosides

The gangliosides are formed when glycosphingolipids join with N-acetyl neuraminic acid or sialic acid. These are also present in the brain and nervous tissue. Other compound lipids contain sulphate.

Niemann Pick's disease (deficiency of sphingomyelinase due to which sphingomyelins accumulate). Tay Sack's disease (deficiency of hexosaminidase because of which ganglioside accumulates) and Gauchers disease (deficiency of beta glucosidase due to which glucocerebroside accumulates) are due to defective degradation of these complex lipids in the brain.

chapter 4

ENZYMES

The enzymes, with a few exceptions, are proteins with specific catalytic properties. They are present in the living organisms and bring about various biochemical reactions. The enzymes reduce the energy of activation and hence aide in the reaction to take place at a comfortable temperature of 37-38 degree centigrade and a pH of 7.

The enzymes possess an active site or a substrate binding site, to which the substrate binds to form an ES complex which then produces the product and regenerates the enzyme without any alteration. The enzymes exhibit high specificity as the active site can accommodate only specific substrate. For example, hexokinase acts on hexoses while glucokinase is more specific for glucose. The protease acts on proteins, peptidase on peptides, lipase on lipids, amylase on starch and so on. Arginase acts on arginine and glutaminase on glutamine. The arginase and glutaminase have absolute specificity while the protease, peptidase, lipase and amylase have group specificity. Certain enzymes are specific for D or L isomers or cis or trans isomers. For example, the D isomer glucose is metabolised by the enzymes of our body and in the case of amino acids it is the L isomer that is used and in the case of unsaturated fatty acids it is the cis isomer that is metabolised. Further, there is requirement of specific coenzymes for several enzymes, for example, the lactate dehydrogenase needs NAD and succinate dehydrogenase has a requirement of FAD, although both enzymes are involved in the removal of hydrogen from the substrates. The transminases use pyridoxal phosphate, carboxylases use biotin and TPP is required for transketolase reaction and so on.

Initially, the enzymes were named without much scientific basis. The salivary amylase was named as ptyalin and glucose fermenting enzymes were named as invertase. These names fail to convey the name of the substrate or the type of the reaction catalysed. Hence, the need of the hour was to give a unique name and number to each enzyme so as to avoid any confusion in indicating an enzyme, especially when the number of enzymes discovered became very large. One way of naming an enzyme is to write the name of the substrate of the enzyme and add "ase" to it. This way we have amylase acting on amylum or starch, lipase acting on lipids, protease acting on proteins and arginase acting on arginine, etc. This method of naming of enzyme does not provide information about the type of reaction catalysed. Therefore the international union of biochemists (IUB) devised a more scientific way of naming an enzyme in which the lactate dehydrogenase is written as lactate: NAD oxidoreductase, which specifies the substrate as lactate, the coenzyme as NAD and the type of reaction as dehydrogenation (oxidoreduction). Another example is that of glucokinase which is better named as glucose: ATP phosphotransferase, specifying the substrates as glucose and ATP, and the reaction as a transfer of phosphate.

Classification of Enzymes

The enzymes are classified into 6 classes as per the IUB guidelines.

1. Oxidoreductases: These are the enzymes catalysing oxidoreductions in the body. The oxidations in the body are mostly dehydrogenations and either NAD, NADP, FAD or FMN (specificity exists) are used as the hydrogen

acceptors or coenzymes. The examples are: alcohol dehydrogenase, lactate dehydrogenase, succinate dehydrogenase, etc.

2. Transferases: These enzymes transfer various groups like methyl, amino, phosphate, glucose, etc. from one substrate (donor) to another (recipient) and these enzymes are respectively called as methyl transferases, aminotransferases, phosphotransferases and glucan transferases. Some examples are the homocysteine methyl transferase, alanine amino transferase, hexokinase, etc.
3. Hydrolases: They are enzymes which breakdown macromolecules like proteins, lipids and starch into small moles like amino acids, fatty acids and glycerol, and glucose respectively by the addition of water. All the digestive enzymes, protease, lipase and amylase are examples of hydrolases.
4. The lyases also cleave a larger molecule to smaller ones, but without the addition of water. An example of a lyase is ATP: citrate lyase, a cytoplasmic enzyme, that cleaves citrate into acetyl CoA and oxaloacetate, in presence of ATP.
5. The isomerases bring about the interchange of isomers. One example is phosphohexose-isomerse which brings about the interchange of fructose-6 phosphate and glucose-6 phosphate. Another example is phosphotriose isomerase involved in the interchange of glyceraldehyde-3 phosphate and dihydroxyacetone phosphate. The enzyme converting cis-retinene to trans-retinene and back also belongs to this class.
6. The ligases join two substrates, for example, DNA ligases joining two fragments of DNA. Other examples are pyruvate carboxylase joining pyruvate and CO_2 to form oxaloacetate and citrate synthase joining oxaloacetate and acetyl CoA to form citrate.

As per the guidelines of IUB, each enzyme is designated by a 4-digit number, the 1st digit indicates the class of the enzyme, the 2nd the subclass, the third the sub-subclass and 4th digit is the number of the enzyme in the list.

Factors affecting the enzyme reaction

The factors affecting the enzyme action include pH, temperature, concentration of substrate, concentration of enzyme, inhibitors, time and light.

pH of the medium

Each enzyme has an optimum pH at which the activity is maximum. The activity or the velocity of the enzyme can be measured by the moles of the product formed per minute time under optimum pH and temperature. This can be done by taking the substrate and the enzyme in constant quantities in test tubes maintained at different pH. The velocity of the enzyme action is measured in different test tubes maintained at varying pH to arrive at the pH at which the activity is maximum, which is the optimum pH of that enzyme. When the enzyme activity is plotted against pH values, a bell shaped curve is obtained.

Most of the enzymes have an optimum pH around 7, the neutral pH. There are enzymes like

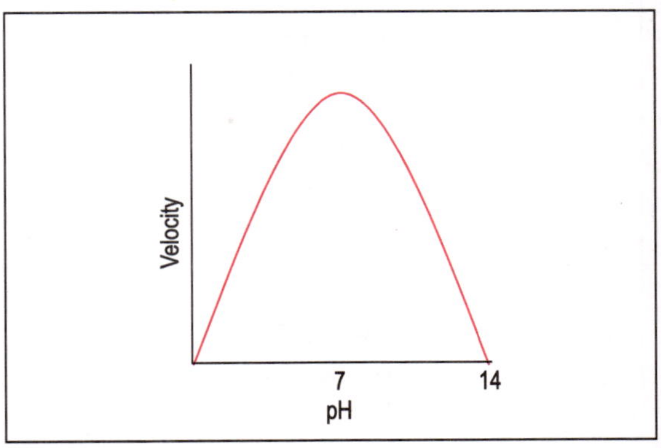

Fig. 4.1: Effects of pH on enzyme activity

pepsin and acid phosphatase with an optimum pH in the acidic range. Enzymes, having an optimum pH in the alkaline range are also present, for example, trypsin, chymotrypsin and alkaline phosphatase are maximally active in an alkaline pH.

Temperature

The temperature is another factor which is important in enzyme action. The optimum temperature for the majority of enzymes is in the range of 37-38 degree centigrade, although, there are enzymes of bacteria inhabiting hot water springs having enzyme action at high temperatures. The optimum temperature of enzyme can be found out by measuring the enzyme activity at different temperatures, which also gives a bell shaped curve, when velocity of enzyme action is plotted against temperature. The enzyme activity is minimal at lower temperatures and increases gradually with an increase in temperature till 38 degree centigrade. This is the result of the increased movement of the substrate molecules, with an increase of temperature, leading to more frequent collision between molecules which results in increased reaction rate. After 40 degree centigrade the activity shows a downward trend as the enzyme gets denatured with a further rise in temperature.

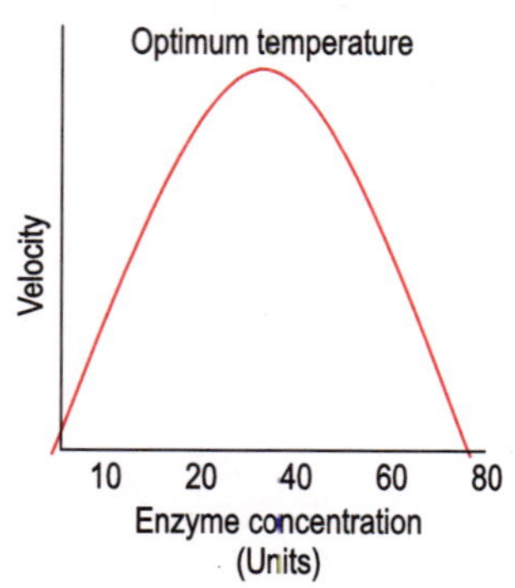

Fig. 4.2 : Effects of temperature enzyme on activity

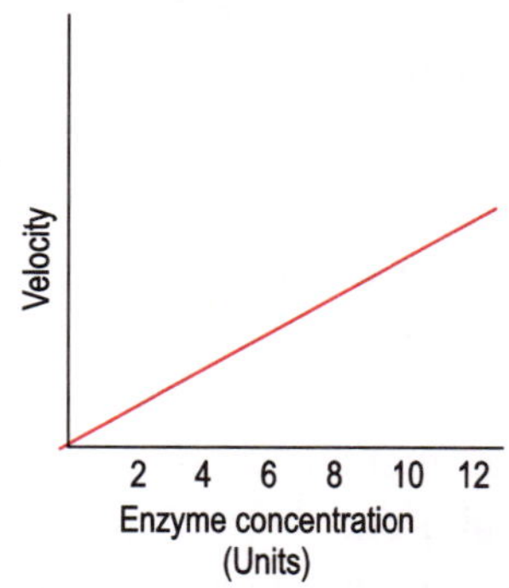

Fig. 4.3 : Effects of enzyme concentration on velocity of reaction

Concentration of substrate

When the concentration of the substrate is varied, initially, the enzyme activity shows a rise, attaining maximal velocity, after which increasing the substrate concentration will not further increase the velocity, thereby giving a hyperbolic curve. In other words, after reaching maximal velocity, the curve straightens due to the saturation of all the enzyme moles with substrate and there are no more enzyme molecules left to combine with the increased concentration of substrate molecules.

The maximum velocity attained is known as V_{max} and the substrate concentration at which

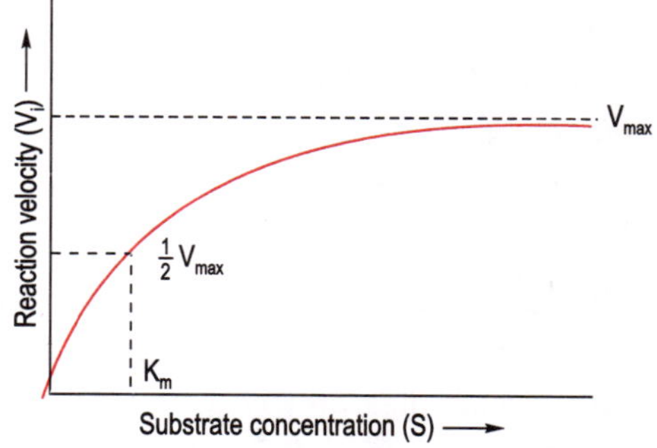

Fig. 4.4 : Effects of substrate concentration on enzyme velocity

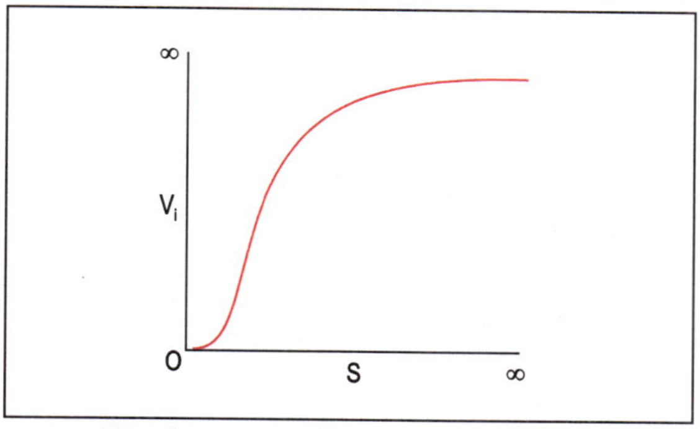

Fig. 4.5 Effects of substrate concentration on enzyme velocity of allosteric enzymes

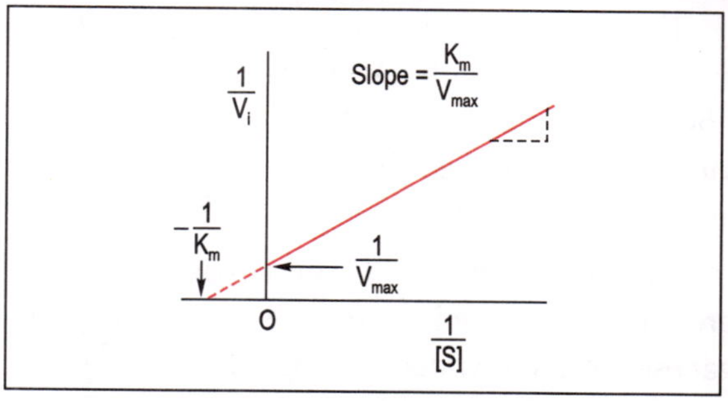

Fig. 4.6 Double reciprocal plot to illustrate the effect of substrate concentration on velocity of allosteric enzymes

there is half maximal velocity is known as Michaelis Menton (K_m) constant for a particular enzyme. The K_m value is always a constant for a particular enzyme. It is like the signature of the enzyme. In the case of allosteric enzymes the velocity versus substrate curve is sigmoid in shape and hence, plotting of substrate concentration with velocity will not be sufficient to calculate the Km. Hence, in this case, the reciprocals of v and s, that is 1/v with 1/s need to be plotted. The K_m can be calculated by extrapolation as shown in the diagram. The value obtained will be $-K_m$.

The competitive inhibitors, structurally similar to the substrates, bind to the active site, thereby competing with the substrate, to occupy the active site of the enzyme, but the inhibitor cannot produce the product, whereas the substrate can. If the concentration of the substrate is increased infinitely, the competitive inhibition can be abolished. Hence, it is reversible. The noncompetitive inhibitors bind with the enzyme at a different site other than the active site. Allosteric inhibition is another type of enzyme inhibition. The enzyme inhibition shall be dealt with in more details elsewhere.

Concentration of enzyme

As the concentration of the enzyme increases, the activity also shows an upward trend in a straight line. This can be demonstrated by varying the concentration of the enzyme in different test tubes and analysing each after adding excess substrate.

Coenzymes

The coenzymes are mostly derived from water soluble vitamins. They are thermostable, dialyzable and form the prosthetic group of some enzymes. The apoenzyme is the protein part and the coenzymes join with it to form the holo enzyme. All the enzymes do not require coenzymes, examples being hydrolases and lyases. The other classes of enzymes need coenzymes for their action. The coenzymes are derived from water soluble vitamins, like thiamine (TPP), riboflavin (FAD and FMN), niacin (NAD and NADP), pyridoxine (pyridoxal phosphate), lipoic acid, biotin, vitamin B_{12} and folic acid.

The NAD is involved in oxidoreduction reactions and the NADPH is required for reductive synthetic steps like fatty acid synthesis, cholesterol and other steroid synthesis. The FAD and FMN are required for dehydrogenation reactions. The biotin is necessary for carboxylation, pyridoxal-phosphate for transamination, vitamin B_{12} and folic acid for one carbon metabolism and thiamin for transketolation. Vitamin C (ascorbic acid) is required for several processes like hydroxylation, for example, in collagen maturation and as a reducing agent.

The role of vitamin K is similar to that of coenzyme. It is necessary for the gamma carboxylation of glutamic acid residues of certain clotting factors and the bone protein osteocalcin, thereby activating these proteins.

The cofactors involved in the action of enzymes are metals like calcium, magnesium, iron, manganese, zinc, molybdenum, copper and selenium.

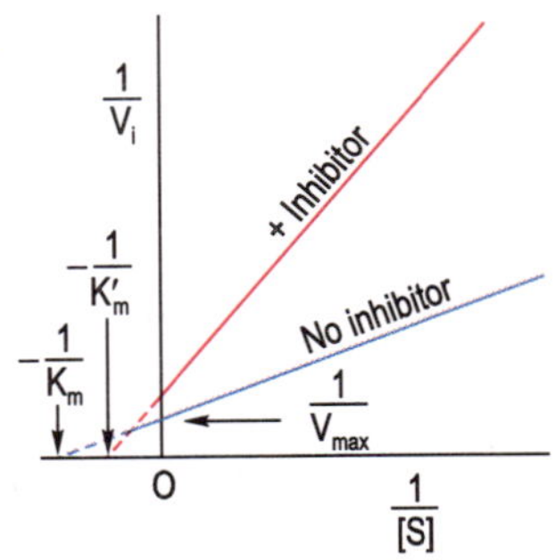

Fig. 4.7: Double reciprocal plot with and without competitive inhibitor for allosteric enzymes

Enzyme Inhibition

The enzyme inhibition is divided into competitive and noncompetitive inhibition.

Competitive inhibition

The competitive inhibitors possess structural resemblance to substrates and hence, can bind at the active site of the enzymes but the enzyme inhibitor complex cannot be converted to the product. The K_m value is increased in the case of competitive inhibition. The competitive inhibition is reversible, meaning, an increase in the substrate can reverse the inhibition. There is no change in the maximum velocity. These can be illustrated by plotting v/s curve in the presence and in the absence of the inhibitor. In the case of allosteric enzymes, double reciprocal plot of 1/v and 1/s is to be drawn. An example of competitive inhibition is that of succinate dehydrogenase by malonate and oxaloacetate as malonate and oxaloacetate are structural analogues of succinate.

The competitive inhibition has several clinical application.

1. The sulpha drugs are analogous to para-aminobenzoic acid (PABA) which has to be combined with pteroyl glutamic acid in bacteria to form folic acid. The folic acid is needed for the synthesis of DNA. As the folic acid synthesis is inhibited, purine and DNA synthesis are stopped. Hence, sulpha drugs are used as antibiotics to kill bacteria. As we humans are unable to synthesise folic acid the sulpha drugs are nontoxic to us.
2. The dicoumarol is an analogue of vitamin K and hence finds its use in the treatment of unnecessary clot formation.
3. The statins are competitive inhibitors of HMG-CoA reductase, the rate limiting enzyme of cholesterol synthesis, and hence used in the treatment of hypercholesterolemia.
4. Physostigmine and neostigmine are employed as inhibitors of choline esterase in the treatment of myasthenia gravis for the prolonged presence of acetyl choline in the synaptosomes.
5. Another example is the use of allopurinol in the treatment of gout. The allopurinol inhibits xanthine oxidase, preventing the formation of uric acid. This is also termed as a suicide inhibition because the xanthine oxidase converts allopurinol to alloxanthine which is a more powerful inhibitor than allopurinol.
6. Warfarin is another competitive inhibitor which prevents blood clotting.
7. Aspirin is an inhibitor of cyclo-oxygenase enzyme which inhibits the formation of thromboxane A_2 helping in the prevention of platelet aggregation. Therefore, aspirin is useful in heart patients.

8. Several anticancer agents like 5-fluorouracil, 5-iodouracil, thioguanine, azaguanine, cytarabine, viderabine, etc. stop DNA replication by action as competitive inhibitors.

Noncompetitive inhibition

They bear no structural resemblance to the substrates as most of them are metals like mercury, lead, arsenic, etc. The iodoacetate is also an example of competitive inhibitor which inhibits the —SH groups of several enzymes. The lead also can combine with the —SH groups present in various enzymes like ferrochelatase thereby causing decreased synthesis of haem. The noncompetitive inhibitors reduce the V_{max} but do not affect the K_m value. In lead poisoning, the British antilewisite is given, because it has several —SH groups to which the lead can be bound and the enzyme —SH groups can be freed from the "clutches" of lead.

There are allosteric inhibitions, in which, the effectors combine with a site other than active site and produce its effect on the active site, so that the reaction is inhibited. There are both positive and negative modulators, the binding of the former leading to activation and that of the latter resulting in inhibition. The binding of allosteric effectors brings about conformational changes on the active site so that the enzyme gets active (if positive modulators are bound) or become inactive (if negative modulators are bound). For example, the enzyme phosphofructokinase, one of the rate limiting enzymes of glycolysis, is inhibited by ATP and activated by ADP and AMP. Another instance is that of pyruvate carboxylase which is allosterically activated by acetyl CoA.

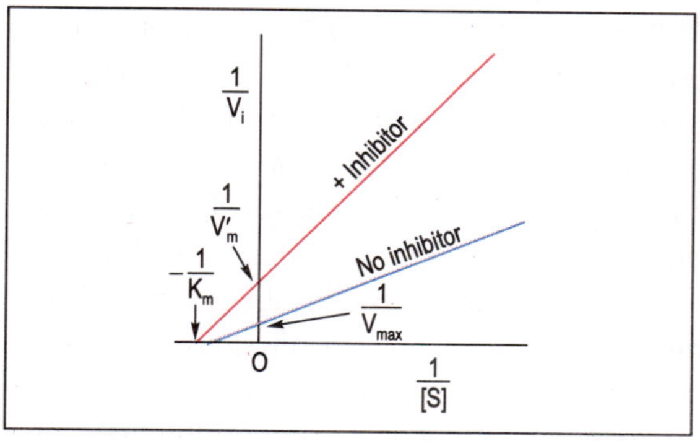

Fig. 4.8: Double reciprocal plot with and without noncompetitive inhibitor for allosteric enzymes

Diagnostic significance of serum enzyme estimation

In this context, the plasma enzymes can be grouped into plasma specific (plasma functional) and plasma nonspecific (plasma nonfunctional) enzymes.

The plasma specific enzymes are those which are always present in plasma in small quantities and these enzymes have specific function in the plasma. The examples are the inactive forms of the enzymes of the blood coagulation system. Another example is pseudocholine esterase.

The plasma nonspecific enzymes are either absent or present in very small amounts under normal healthy state in the plasma, but, increases several folds due to a disease process. The examples are alanine aminotransferase (ALT) in hepatocellular jaundice and CK MB (Creatine Kinase–MB) in acute myocardial infarction (AMI). These marker enzymes are released into the plasma from the tissues due to tissue damage resulting from a disease process. The enzymes coming out of the damaged cells leak into the plasma leading to their raised levels in the serum. The larger the rise in the plasma level of the enzyme, the greater is the amount of the tissue damaged. The cytoplasmic enzymes appear early in the plasma, followed by the mitochondrial and other enzymes present in the other subcellular particles.

Serum enzyme levels in various types of jaundice

The hepatocellular jaundice can be diagnosed by estimating the serum levels of alanine aminotransferase (ALT/GPT), aspartate aminotransferase (AST/GOT) and alkaline phosphatase (ALP) in addition to the estimation of total, direct and indirect bilirubin. The normal range of serum ALT is below 30 U/litre and an increase in ALT is indicative of hepatocellular jaundice. The ALT increase can be before bilirubin elevation. In severe cases of jaundice, the ALT levels increase by several hundreds of folds. The aspartate amino transferase (AST) also shows an increase but its increase is less than that of ALT showing that ALT is a better marker of hepatitis. The normal range of serum AST is less than 20 U/litre. A sudden fall in these aminotransferases after a major rise indicates poor prognosis for the patient.

The level of alkaline phosphatase (the normal value being less than 40 U/litre) increases moderately in hepatocellular jaundice, although greater elevation of ALP takes place in obstructive jaundice as the bile duct has higher concentration of ALP than the hepatocytes. In most cases of obstructive jaundice there is no increase of ALT and AST. So is the case with haemolytic jaundice as well, where none of the enzymes, ALT, AST or ALP are elevated. Thus, the differential diagnosis of jaundice can be made by assaying the enzyme levels coupled with bilirubin estimation.

The gamma glutamyl transferase is an indicator of alcoholic cirrhosis of the liver.

Isoenzymes

Isoenzymes are multiple molecular forms of the same enzyme or they are physically distinct forms of the same enzyme activity. The isoenzymes have slightly altered amino acid sequences and so can be separated by electrophoresis. Further, the isoenzymes are produced by different tissues and have different quaternary structures.

The clinically important isoenzymes include lactate dehydrogenase (LDH), creatine kinase (CK) and alkaline phosphatase.

The LDH is a tetramer having 2 types of polypeptide chains H and/or M. Different forms of LDH isoenzymes are present in different tissues like heart, RBC, brain, liver and skeletal muscle. Each is different from one another in the proportion of H and M polypeptide chains. Hence, the heart form has 4H chains, the RBC has 3H and 1M chains, the brain has 2H and 2M chains, the liver has 1H and 3M chains and skeletal muscle form has 4M chains. They can be separated by electrophoresis. The LDH-1 is increased in heart disease, LDH-4 in liver disease and LDH-5 in muscle trauma. Normally, LDH-2 is more than LDH-1, but in acute myocardial infarction, the LDH-1 becomes more the LDH-2, this being called as flipped pattern.

Another isoenzyme of clinical importance is creatine kinase (CK) or creatine phosphokinase. The CK has three isoenzymes, namely, CK-MM, CK-MB and CK-BB, present in skeletal muscle, heart and brain respectively. Out of the total CK present in the normal serum, approximately, 90 per cent is contributed by CK-MM, 9 per cent by CK-MB and 1 per cent is contributed by CK-BB. Hence, an increase of CK-MB is indicative of AMI. The CK-MB is an early cardiac marker as it gets elevated within 4 to 6 hours after the onset of AMI/chest pain. The CK-MM increase points towards muscular dystrophy.

Alkaline phosphatase has several isoforms. One type is produced in the hepatocytes, another in the walls of the bile duct, a third type in the bones, a fourth type is in the placenta, fifth type in the leukocytes and another type in the small intestine. Hence, the disease in any of these tissues can lead to the increase of the specific isoenzyme produced by it, and hence, is of diagnostic importance. The maximum rise in ALP is seen in bone disease. The ALP shows a physiological rise in pregnancy and in growing children. The levels are increased in both hepatitis and obstructive jaundice, the increase being more in the latter. The ALP is increased in ulcerative colitis and also in leukaemias. The haemolytic jaundice does not show any increase in serum ALP.

Serum enzyme levels in AMI

The estimation of serum enzyme levels has great importance in diagnosing AMI, when the ECG changes are not conclusive, especially, for the early diagnosis of AMI and microinfarcts. Out of the enzyme markers, CK-MB, AST, ALT and LDH-1 are important. The rise of AST is more than of ALT, which is in contrast to liver disease where the opposite condition prevails. The rise of LDH, AST and ALT occurs about 20 hours after the infraction has taken place. Compared to these enzymes, the CK-MB rises within 4-6 hours of AMI, and hence, is of great importance as an early marker of AMI.

The cardiac troponins, T and I, although they are not enzymes, are other early marker proteins, employed frequently to diagnose AMI. These proteins, show an increase within 6 hours of the infarction. The cardiac muscle and skeletal muscle forms of troponins are different isoforms with different amino acid sequences, and hence, can be estimated separately.

So the levels of serum CK-MB and cardiac troponins, T and I, serve to arrive at a diagnosis of AMI along with ECG changes, early enough to start treatment.

Other markers of myocardial infarction include ischaemia modified albumin, myeloperoxidase and the isoenzyme of glycogen phosphorylase BB.

Other enzymes of diagnostic value

The other enzymes/proteins, routinely investigated in clinical laboratories, include amylase, lipase, glucose-6 phosphate dehydrogenase, acid phosphatase, PSA and brain natriuretic peptides.

The serum amylase estimation is important in salivary gland disease, which shows increased concentrations of amylase. The amylase is also elevated in pancreatitis. The lipase also is elevated in pancreatitis and hence, the differential diagnosis of salivary gland pancreatic disease can be made.

The glucose 6-phosphate dehydrogenase deficiency leads to haemolytic anaemia and jaundice. The G6PD deficiency results in less production of NADPH which is required to form reduced glutathione from the oxidised one by glutathione reductase.

The oxidised form of glutathione is produced when the H_2O_2 is detoxified to H_2O by glutathione peroxidase. Hence, the G6PD levels are estimated.

The brain natriuretic peptide (BNP) and atrial natriuretic peptide (ANP) are of clinical importance. The ANP as the name is indicative is formed in the atria of the heart. The brain and the cardiac ventricles have the pro-BNP which is cleaved to form an active BNP consisting the C terminal portion and the inactive N terminal region. The natriuretic peptides prevent salt and water loss from the body. The patients with congestive cardiac failure have higher concentrations of ANP and BNP. The patients with chronic obstructive pulmonary obstruction have elevated levels of BNP.

The acid phosphatase levels are increased in prostate cancer. Another marker for prostate cancer is prostate specific antigen (PSA).

Enzymes used as therapeutic agents

Pepsin and trypsin are given to patients with digestive problems. Streptokinase and urokinase find their use in myocardial infarction to dissolve the fibrin clots. The asparginase is an anticancer agent. The ADA gene is used for the treatment of severe combined immunodeficiency (SCID).

Enzymes as reagents

Enzymes like restriction endonuclease, glucose oxidase, peroxidase, urease, cholesterol oxidase, esterase, etc. are commonly being used as reagents in the clinical labs.

chapter 5

Plasma Proteins

The plasma can be prepared when the blood is collected in a bottle/tube containing an anticoagulant like heparin or EDTA or oxalate. The centrifugation of such blood forms a supernatant plasma and a pack of cells below. The serum, on the other hand, is obtained by collecting the blood in a container having no anticoagulant when the blood will clot after keeping for about half an hour. The only difference between plasma and serum is that the plasma contains fibrinogen, whereas the serum is devoid of it as the fibrinogen is consumed in the clot. For all the other constituents, the plasma and serum are practically the same.

The total plasma protein concentration is 6 to 8 g/dl. The plasma proteins can be separated by electrophoresis, which is the movement of charged particles in an electric field. The electrophoresis can be done on a medium like Whatmann filter paper or on agar/agarose gel or on cellulose acetate paper. The electrophoresis is to be carried in a buffer with constant pH. The buffer used is usually barbiturate buffer at a pH of 8.6. It is a mixture of sodium barbiturate and barbituric acid. The plasma/serum is charged on the medium and placed in the electrophoretic chamber containing the buffer, in which the negative and positive electrodes are kept. A constant current is applied for about 50 minutes in the case of gel/cellulose acetate, and 10 hours in case of Whatmann paper. As the medium is alkaline the proteins become negatively charged and so migrate towards anode. The charging of the sample is done on the negative electrode. The extent of migration of the different fractions of proteins depends on the electrical charge of the individual fractions, the maximum electronegative albumin runs faster and farther to the anode, the other fractions distribute themselves as per the strength of their charges. The least electronegative are the gamma globulins which migrate the least.

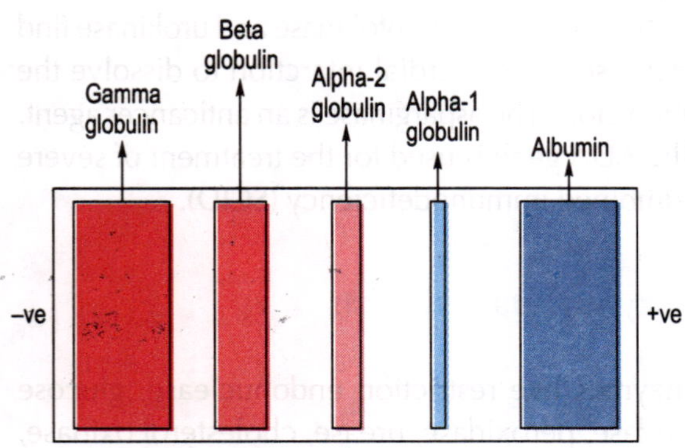

Fig. 5.1: Plasma protein electrophoresis

Hence, the electrophoresis separates the plasma proteins into fractions like albumin, alpha-1, alpha-2, beta, and gamma globulins, arranged from anode to cathode. After the run is over, it is dried and stained with a protein stain like amido black when the blue coloured protein bands become visible. The excess of stain is washed away with 10 per cent acetic acid, when the protein bands become clearer. The densitometers can be used to measure the intensity of these bands which is directly proportional to their concentrations.

Plasma albumin

Out of all the plasma proteins albumin has the highest in concentration, the normal range being 3.5 to 5.5g/dl. The liver is the organ which synthesises albumin and all the other plasma proteins except immunoglobulins. About 12 grams of albumin is synthesised everyday. The half-life of albumin is 20 days. It is made up of a single polypeptide chain and consists of about 458 amino acids and has a molecular weight of 69000, being smaller than the globulins with a very large molecular weight.

Being a small protein molecule, but, high in concentration, the albumin exerts a high osmotic pressure and hence, is involved in the fluid exchange between intravascular extracellular compartments. The small molecules like glucose, urea and electrolytes, etc. can go and come freely between the compartments. The large molecules like proteins are not able to do this because they are not permeable across the membranes. The osmotic pressure due to the small molecules is the same both in the intravascular and extravascular compartments and hence, cancels out. Therefore, the fluid exchange is the result of the blood pressure generated by the working of the heart acting from inside to outside of the vessels and the osmotic pressure due to the plasma proteins (oncotic pressure) acting from outside to inside. Accordingly, at the arterial end of the capillaries, the blood pressure is about 35mm of Hg and the oncotic pressure is about 25mm of Hg, both acting in opposite directions, drives the fluid from the vascular compartment to outside with a net filtration of 10mm of Hg. At the venous end of the capillaries the blood pressure is a low of 15mm of Hg, but, the oncotic pressure remains at 25mm of Hg, as the proteins are impermeable across the vessel wall. This creates a net absorption pressure of 10mm of Hg and the same amount of fluid which had gone out at the arterial end returns at the venous end. This is known as Starling's hypothesis.

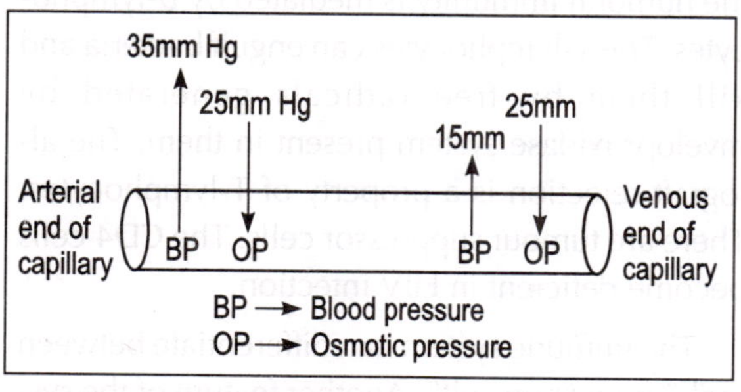

Fig. 5.2: Starling's hypothesis

As the plasma albumin is the major contributor of oncotic pressure, due to its higher concentration and smaller size (the osmotic pressure being dependent on the number of particles and not on their size), a decrease in the albumin lowers the osmotic pressure resulting in the exit of more fluid than the return. This condition ends up in the collection of fluid in the extracellular spaces, being termed as *oedema*.

Low levels of albumin is encountered in nephritic syndrome in which albumin, being the smallest of the plasma proteins, is excreted in the urine leading to low levels of albumin in the plasma and hence, causes oedema. In the case of nephritic syndrome, the alpha-2 macroglobulin being a large protein is not lost in the urine.

Another cause of low albumin is due to its decreased synthesis as in the case of hepatitis and liver cirrhosis. This can be differentiated by electrophoresis of the plasma proteins. The albumin band will be light in both the above pathological states, but the alpha-2 macroglobulin will be more prominent in nephritic syndrome, whereas it will be normal in liver diseases.

Yet another cause of low albumin is protein malnutrition, called as *kwashiorkor*, in small children. Another condition, protein losing enteropathy, can also result in decreased albumin status.

The plasma albumin also has a role as a transport protein. It transports free bilirubin, free fatty acids, drugs like aspirin, sulpha, etc., metals like calcium, copper and steroids. As albumin contains all the essential amino acids it has nutritive function. Another function of albumin is to act as a plasma buffer helping in the maintenance of blood pH.

Alpha-1 antitrypsin

It migrates in the alpha-1 globulin fraction on electrophoresis and is also called as alpha-1 antiproteinase and inhibits proteases. During lung infection, the macrophages present at the site of infection produce elastase enzyme which can destroy the lungs. The alpha-1 antiproteinase inactivates the elastase and prevents damage to the lung tissue. Hence, a deficiency of the alpha-1 antiproteinase may cause emphysema. In smokers, the 368th methionine of the alpha-1 antitrypsin is oxidised to sulphoxide, thereby inactivating it. Therefore, emphysema is common in the smokers. Alpha-1 antitrypsin is one of the acute phase reactants which remains low in normal state, but increases several folds in infectious/inflammatory conditions.

Haptoglobins

The haptoglobins are required to bind to extracorpuscular haemoglobin, produced due to haemolysis. The haemoglobin, being a small protein and having a molecular weight of about 59000, may be lost in the urine, resulting in loss of valuable iron and globin chains. Normally this loss is prevented by the haptoglobin binding with the haemoglobin in the plasma to form a large complex of [haemoglobin-haptoglobin] which cannot be filtered at the glomerulus. The haptoglobin levels are decreased in mismatch blood transfusion due to consumption of more haptoglobin for binding to haemoglobin in the plasma, caused due to haemolysis. In that way, haptoglobins are also considered to be acute phase reactants.

Ceruloplasmin

The ceruloplasmin is a copper binding protein. The apoceruloplasmin binds with copper to form ceruloplasmin. A deficiency of ceruloplasmin causes copper toxicity and causes the Wilson's disease developing a bluish green ring around the cornea, which is known as *Keischer Fleischer ring*.

Transferrin

It is a protein for the transport of iron. Two-third of transferrin is saturated with iron and one-third is free or unsaturated. The apotransferrin binds with iron to form transferrin. Only about 2-3mg of iron are absorbed every day from the diet. The diet contains about 20-25mg of iron, but only 10 per cent of it is absorbed. If more than 2-3mg of iron is absorbed everyday, it can lead to iron overload and toxicity. All the heavy metals in the free form is toxic. The iron if deposited in tissues like the liver or pancreas can damage the corresponding tissues. The serum concentration of iron is about 300mg/dl. The iron storage protein is ferritin. The iron toxicity can cause haemochromatosis or haemosiderosis.

The other plasma proteins include haemopexin (haem binding protein). Transcortin (cortisol binding protein), transthyretin (thyroxine binding protein/pre-albumin), thyroxine binding globulins, retinol binding protein, C-reactive protein, immunoglobulins, blood clotting factors, enzymes and lipoproteins. An increase in C-reactive protein is a risk factor for future cardiovascular problems. The elevated level of high-sensitive (hs) C-reactive protein indicates an inflammatory state.

Immunoglobulins

There are two types of immunity, the cell mediated and the humoral immunities. The cell mediated immunity is mediated by T-lymphocytes and the humoral immunity is mediated by B-lymphocytes. The T-lymphocytes can engulf bacteria and kill them by free radicals generated by myeloproxidase system present in them. The allograft rejection is a property of T-lymphocytes. There are tumour suppressor cells. The CD4 cells become deficient in HIV infection.

The immune system can differentiate between "self" from "non-self". Another feature of the system is that it has memory. If an antigen is injected

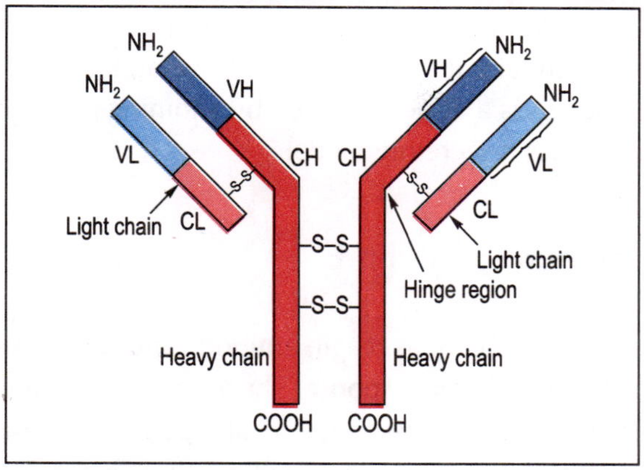

Fig. 5.3: Immunoglobulin structure

into an animal, the animal produces antibodies against the antigen, which takes about 10 days to initiate the process, reaches a peak at 20 days, comes to the baseline after about 30 days. Further, if the antigen is given after a lapse of 6 months, the antibody response will be faster and stronger and persists for longer period. This represents memory of the immune system. The primary response antibody is IgM and the secondary response ones are IgM and IgA.

All the immunoglobulins are synthesised by the lymphocytes, namely the B-lymphocytes. The immunoglobulins, otherwise called as antibodies, are classified into 5 major classes: IgG, IgM, IgA, IgD and IgE.

Structure of Immunoglobulin

The immunoglobulins have a quaternary structure with 2 heavy (H) chains and 2 light (L) chains. They are connected by disulfide bridges made of cysteine amino acids. The amino terminus portions of both light and heavy chains have variable (V) amino acid sequences, and the carboxy terminus regions have constant (C) amino acid sequences. Half of the light chain towards the amino terminus has variable sequence, whereas the one-third of the heavy chain towards the aminoterminus has V sequence. The L chains are of 2 types, the kappa and lambda, of which the kappa is seen more frequently. The H chains are of 5 types: gamma (G), macro (M), alpha (A), delta (D) and epsilon (E) and the immunoglobulin classes derive their names from their corresponding H chains.

Immunoglobulin G

The IgG is the simplest of all the immunoglobulins, structurewise and also the smallest, with two gamma heavy chains and two light chains. Being a relatively small protein, it can pass across the placenta and hence, gives immunity to the foetus and the newborn till about 8-10 months of age, after which the child is able to form the antibodies. It can bind to macrophages and has the property of complement fixation. It has the highest concentration in the plasma. The IgG is a secondary response antibody. The antigen binding region is made up of both L and H chains.

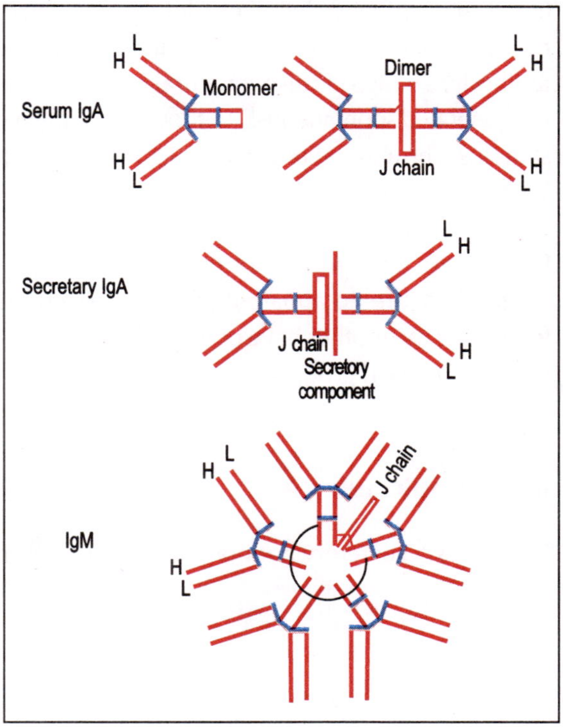

Fig. 5.4: Diagrammatic representation of serum IgA, secretory IgA, and IgM

The Rh-antibodies are IgG type and hence, can cross the placenta. This can be reflected in the condition of erythroblastosis foetalis. When an Rh negative mother has an Rh positive foetus, during the time of delivery, the blood of the foetus goes to the mother and elicits the production of antibodies against the foetal RBCs. Nothing untoward happens to the first child, but when a second Rh positive child is conceived, the Rh antibodies, already generated in the mother during the first pregnancy enters the foetal circulation and causes haemolysis in the foetus which may result in foetal death. The second foetus can be saved by giving anti-Rh antibody injection to the mother at the time of delivery.

Immunoglobulin M

This immunoglobulin is a pentamer consisting of 5 subunits, each with 2H and 2L chains, making IgM pentavalent. It also has 5 antigen binding sites and is more effective in killing blood borne bacteria. As mentioned earlier, it is a primary response antibody. The IgM is also a natural antibody, being generated spontaneously without any antigenic stimulus. The antibodies of the A, B, O blood group system are IgM type which cannot cross placental barrier. They have the property of complement fixation. As the IgM cannot pass across the placenta, there are no apparent problems when the foetus and the mother have different blood groups of A, B, O blood group system unlike the Rh system.

Immunoglobulin A

The IgA is a secretory antibody and is present in the seromucous secretions like respiratory tract, urogenital tract, intestinal tract, sweat, tears, etc. The IgA is a dimer attached by a joining piece. It also has a secretory piece.

Immunoglobulin E

It is also known as reagin antibody and is responsible for allergic response. The immunoglobulin E is produced in response to allergens like dust, pollen grains, mites, penicillin, etc. When penicillin injection is given to those allergic to it, the IgE gets attached to the mast cells which secretes histamine and slow reacting substance of anaphylaxis and causes degranulation of mast cells. This is immediate hypersensitivity reaction or exaggerated immune response. When tuberculin test is carried out, if the tubercle bacteria are present in the sputum, it gives a reddish patch around the site of the test within 48 to 72 hours, which is known as delayed hypersensitivity reaction.

The function of the fifth immunoglobulin, the IgD is not known.

The hypergamma globulinaemia can be detected in chronic infections like TB, leprosy, malaria, etc. This can be seen as a thick band of gamma globulins in the electrophoretogram. In multiple myeloma, a sharp band in the gamma region can be seen. This is called as M band. This is the result of the proliferation of a single clone of immunoglobulins, may be IgM or IgA. Such antibodies are of little use to the body as only one type of immunoglobulins are generated. Another type of immune disorder is Waldenstrom's macroglobinemia.

chapter 6

Haemoglobin, Myoglobin and Collagen

HAEMOGLOBIN, MYOGLOBIN AND COLLAGEN: STRUCTURE AND FUNCTIONS

Haemoglobin

The major haemoglobin present in adults, the adult haemoglobin (HbA), is composed of 4 polypeptide chains, 2 alpha and 2 beta chains. The foetal haemoglobin (HbF) is made of 2 alpha and 2 gamma chains. The haemoglobin has a quaternary structure. The HbF has greater affinity for oxygen than HbA and this helps the foetus to extract oxygen from the maternal blood. After birth, the HbF is not suitable because it does not release oxygen easily which may lead to hypoxia.

Each of the globin chain is linked to a haem moiety which is a tetrapyrrole (porphyrine) having iron in the centre. The 4 valences of iron are attached to the nitrogens of each pyrrole ring, the 5th valance to the proximal histidine (F_8) and the 6th valence is loosely bound to the oxygen. Lying just below the oxygen molecule in the oxygenated haemoglobin is the distal histidine (E_7). Due to the presence of this histidine, the second atom of oxygen is bound at an angle with the plane of the haem, whereas the first atom of oxygen is perpendicular to the haem plane. Because of oxygen binding at an angle, it becomes a weak binding, and so can release the oxygen easily in the tissues. The haemoglobin binds to oxygen in the lungs where the O_2 is very high, at the same time releasing CO_2, which is expelled out. The reverse takes place in the tissues where the oxygen is released and CO_2 is bound to the Hb. This effect is known as Bohr's effect. An increase in acidity and CO_2 helps in the release of oxygen, whereas an increase in O_2 and a decrease in acidity expels the CO_2. The 2, 3-BPG, an intermediate in RBC glycolysis, functions in the unloading of oxygen in the tissues.

The alpha chain of haemoglobin has 145 amino acids and the beta chain has 141 amino acids. Each chain has 8 helices marked from A through H. The haem residue is lying in a crevice between helices E and F. The iron lies 0.03 micrometers above the place of the porphyrin in the deoxygenated state. On oxygenation, the iron is brought to the plane of the porphyrin ring. With this, the polypeptide chain linked to the iron through histidine is also shifted which in turn alters the conformation of haemoglobin molecule which increases the affinity of oxygen binding. The deoxygenated form is called as the 'T' (taut) form and the oxygenated form is the 'R' (relaxed) form.

Fig. 6.1: Structure of haemoglobin

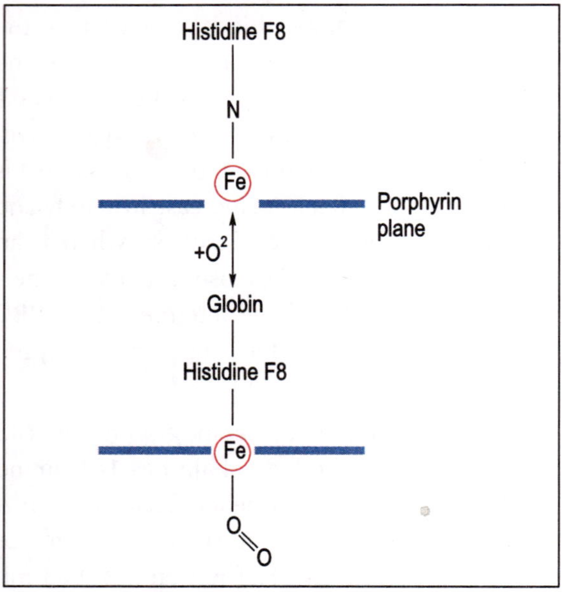

Fig. 6.2: **Oxygen binding to haemoglobin**

The oxygen dissociation curve of haemoglobin is sigmoid or S-shaped, whereas that of myoglobin curve is hyperbolic. The binding of one molecule of oxygen to one polypeptide chain of haemoglobin increases the affinity of oxygen binding to the second polypeptide chain, which in turn further increases the affinity of oxygen binding to other polypeptide chains. Due to this, initially, the oxygen binding is slower, but after a few moles of oxygen are bound to the haemoglobin, the affinity of further oxygen binding is increased. That is why the oxygen dissociation/association curve of haemoglobin is sigmoid in nature. This is also known as cooperative binding, haem-haem interaction or allosteric effect. The same phenomena is seen in the tissues in which, the release of one molecule of oxygen from one polypeptide chain of haemoglobin helps in the release of the other three moles of oxygen.

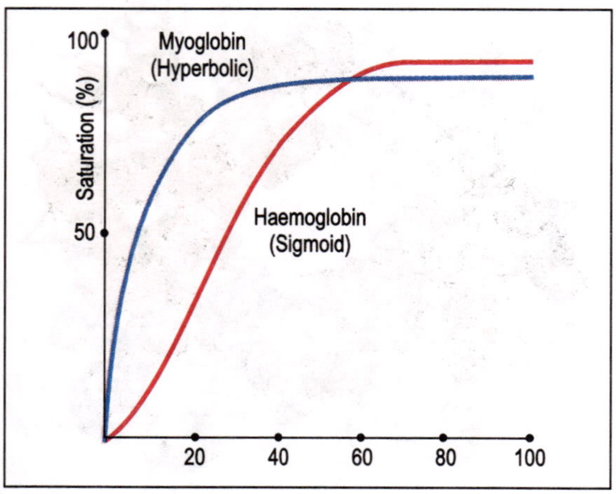

Fig. 6.3: **Oxygen dissociation curve of haemoglobin and myoglobin**

The 2, 3-bis phosphoglycerate, an intermediate in the glycolysis in erythrocytes, forms salt bridges with the basic amino acids of haemoglobin which decreases the affinity of oxygen binding leading to the unloading of oxygen. This is the taut form of haemoglobin. The relaxed form of haemoglobin has no salt bridges. The foetal haemoglobin has more affinity for oxygen as 2, 3-BPG has less effect on HbF. The 2, 3-BPG concentration is found to be increased in high altitudes as an adaptive mechanism.

The carbon monoxide binds with haemoglobin with greater affinity than oxygen and cannot be released leading to CO poisoning.

Myoglobin

The myoglobin is present in muscles for storing oxygen. It cannot function as a transporter of oxygen as it is not able to release oxygen at a partial pressure of O_2 around 30-40mm of Hg, usually encountered in the actively metabolising tissues. The myoglobin liberates oxygen only at a very low partial pressure of O_2 of below 10mm of Hg, which is obtained only in muscles performing heavy exercise.

It has only one polypeptide chain having 141 amino acids. It is a protein with tertiary structure in comparison to the haemoglobin possessing a quarternary level of structure. The myoglobin can bind only one molecule of oxygen, whereas the haemoglobin can bind 4 moles of oxygen. The myoglobin has no allosteric properties and does not exhibit cooperative binding or haem-haem interaction. As the myoglobin remains saturated (80 per cent) with oxygen even at a partial pressure of oxygen of 10mm of Hg, it is not suitable to carry oxygen, but can serve as a storage protein for oxygen.

The myoglobin used to find application as a diagnostic marker of acute myocardial infarction, in the past, but no more now, as it is degraded very quickly. Further, with the advent of better cardiac markers like CK-MB and cardiac troponins, the diagnostic use of myoglobin has decreased over the years.

Collagen

The collagen, elastin, fibronectin, keratin, etc. are all structural proteins of the body. The collagen is present in the extracellular matrix and it keeps all the cells of a tissue together. It is a heterogeneous group of proteins present in bones, cartilages, tendons, walls of the blood vessels, skin and lens of the eyes. It is a protein with a very long half-life and is seldom turned over, thereby attaining the status of permanent proteins. The collagen synthesis starting early in life lasts the whole life. The collagen consists of three polypeptide chains, twisted over one another and made strong by cross-linking the protein chains.

The synthesis of collagen takes place in the fibroblasts and the collagen formed is called as tropocollagen or procollagen which consists of 3 polypeptide chains twisted over one another. Each polypeptide chain contains around one thousand amino acids, every third amino acid being glycine. Majority of the remaining amino acids are made up by lysine, proline, hydroxy lysine and hydroxy proline. As the collagen is an extracellular protein, its synthesis takes place in the ribosomes attached to the endoplasmic reticulum. The newly formed collagen is immature and is soluble and has less tensile strength. This has to undergo posttranslational modifications like glycosylation (in the golgi bodies) during its passage through the lumen of the ER. Further changes required are the hydroxylation of amino acids lysine and proline to hydroxyl lysine and hydroxyl proline respectively. These hydroxylation reactions are catalysed by lysyl hydroxylase and prolyl hydroxylase for lysine and proline respectively. These enzymes have a requirement of ferrous form of iron and to keep the iron in the reduced state vitamin C is necessary. That is why in the absence of ascorbic acid survey is prevalent due to defect in the maturation of collagen. The hydroxy lysine is oxidised by lysyl oxidase and hydroxyproline is oxidised by prolyl oxidase, both of these enzymes require copper. Cross-linkages are formed between the oxidised lysine and oxidised proline on different polypeptide chains to make the collagen molecule an insoluble and strong fibrous protein with great tensile strength. After secretion into the extracellular space, certain peptidases remove certain amino acids from both the amino terminus and the carboxyl terminus to form the mature collagen.

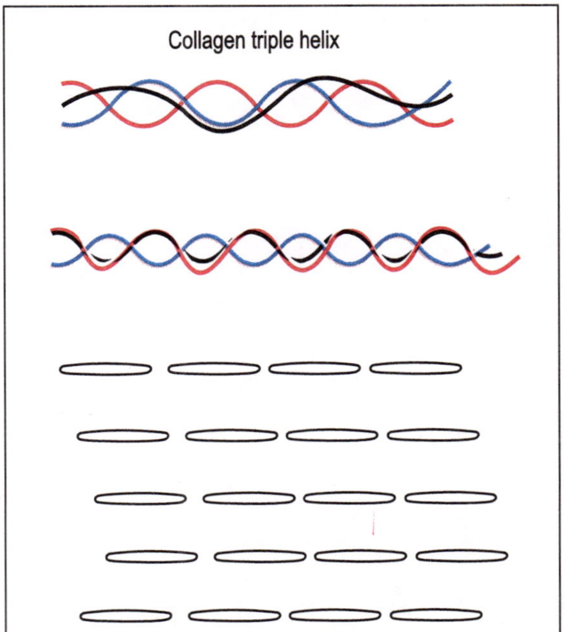

Fig. 6.4: Diagram of quarter staggered arrangement of collagen

The collagen molecule is arranged in a quarter- staggered triple helix with cross-linkages which contributes to its tensile strength. The enzyme which degrades collagen is collagenase present in the soil bacterium clostridium histolyticum. The collagenase is absent in the humans. The gas gangrene seen in uncontrolled diabetes mellitus is due to the action of the collagenase. The continuous cross-linking taking place in the collagen during the lifespan of an individual can explain the lack of elasticity of the blood vessels and the skin, and the complications arising out of it.

Collagen disorders

The examples of collagen disorders include the following:

1. Osteogenesis imperfecta
2. Dermatoparaxis
3. Marfan's syndrome
4. Lathyrism
5. Ehlers-Danlos syndrome

All these are characterised by skeletal deformities, loose joints and skin. The blood vessels become less elastic. In all of the above, the quarter staggered structure of collagen is disturbed resulting in immature collagen.

In Osteogenesis imperfecta certain glycine molecules are replaced. In such children even minor fall can lead to fractures in the bones.

chapter 7

Haem Synthesis, Porphyrias, Haem Breakdown and Jaundice

The synthesis of haem starts with succinyl CoA and glycine which condense to form alpha-amino beta-keto adipic acid in presence of aminolevulinic acid synthase (ALA synthase). In this reaction pyridoxal phosphate is needed as a coenzyme and CoA-SH is removed. The next reaction catalysed by the same enzyme removes one CO_2 molecule and produces delta-amino levulinic acid, the reaction requiring pyridoxal phosphate. The ALA synthase is the rate limiting enzyme of the pathway which is regulated by feedback inhibition. The ALA synthase is located inside the mitochondria.

The next step involves the formation of porphobilinogen (PBG), taking place in the cytoplasm. It is catalysed by ALA dehydratase, having a requirement of zinc, leads to the condensation of two moles of ALA forming PBG, with the liberation of 2 molecules of water.

The third reaction is catalysed by PBG deaminase, alternatively named as uroporphyrinogen UPG-1 synthase, and UPG-3 cosynthase condenses 4 molecules of PBG, liberating 4 molecules of ammonia, to form UPG-III which is the naturally seen porphyrinogen. It is worth mentioning that the porphyrinogens are colourless compounds, whereas the oxidised porphyrins are coloured.

The 4th step, catalysed by UPG decarboxylase, is to convert UPG-III to coproporphyrinogen CPG-III, with the liberation of 4 moles of CO_2.

The next 3 reactions occur in the mitochondria. The CPG-oxidase in the 5th reaction forms protoporphyrinogen PPG-III from CPG-III, the reaction needing oxygen.

The 6th step generates protoporphyrin (PP)-III by oxidising PPG-III, catalysed by PPG oxidase, requiring molecular oxygen. The protoporphyrins are coloured which have methenyl bridges.

The 7th step is catalysed by ferrochelatase or haem synthase which catalyses the incorporation of ferrous iron into protoporphyrin III to generate heme.

Regulation of haem synthesis

The haem synthesis is regulated by feedback inhibition. If the haem is in excess, it will allosterically inhibit the ALA synthase enzyme.

The regulation of haem is achieved by repression derepression mechanism also. When the haem is in sufficient quantities, acting as a core-

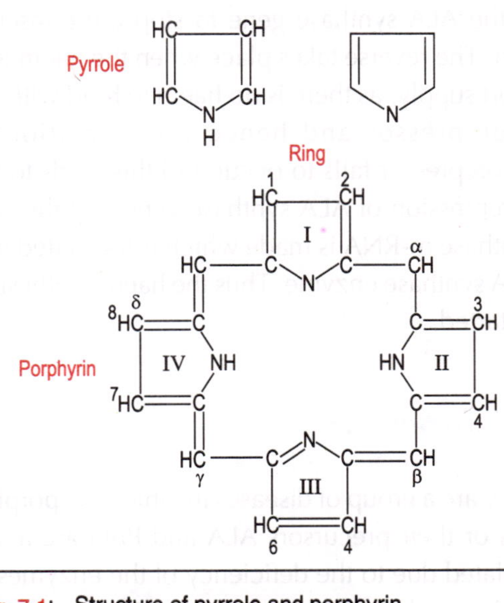

Fig. 7.1: Structure of pyrrole and porphyrin

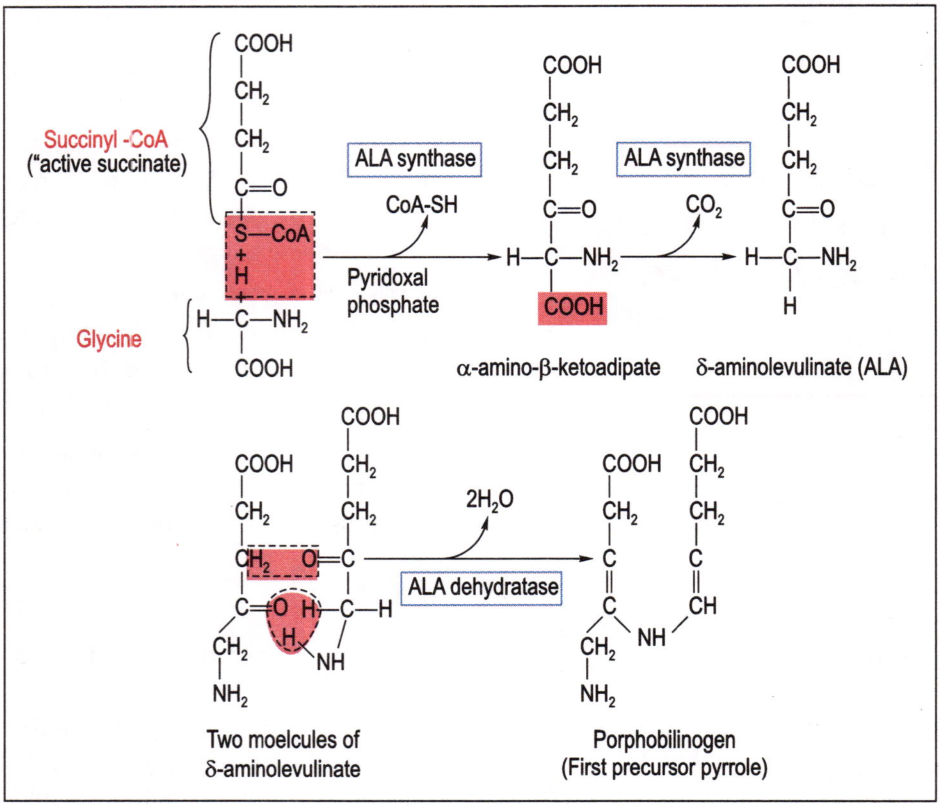

Fig. 7.2: Initial steps in haem synthesis

pressor combines with the aporepressor to form holorepressor, which binds at the promoter region of the ALA synthase gene to stop the transcription. The reverse takes place when the haem is in short supply, as there is no haem to bind with the aporepressor and hence the formation of holorepressor fails to occur and this leads to the derepression of ALA synthase gene and the ALA synthase m-RNA is made which is translated into ALA synthase enzyme. Thus the haem synthesis is restarted.

Porphyrias

They are a group of diseases in which the porphyrins or their precursors ALA and PBG are accumulated due to the deficiency of the enzymes of the biosynthetic pathway of haem. The porphyrias belong to three categories (a) hepatic (b) erythropoietic and (c) erythrohepatic.

The porphyrias are also classified based on the site of enzyme deficiency and the type of substrates accumulated. They are mostly autosomal dominant traits. Associated anemia is not a feature of porphyria.

Acute intermittent porphyria

This is due to the deficiency of PBG deaminase or UPG-I synthase. As the synthesis of haem does not occur the ALA synthase is derepressed, and so, the increased activity of this enzyme results in elevated levels of ALA and PBG which accumulates in the brain causing neurological problems. They can also cause abdominal pain due to the accumulation of ALA and PBG in the abdominal nerves. As the porphyrins are not accumulated there is no cutaneous photosensitivity.

The barbiturates are found to induce the ALA synthase and can aggravate AIP. Starvation worsens porphyria, whereas glucose can improve the condition. Postmenopausal women have more severe symptoms.

The other types of porphyrias include congenital erythropoietic porphyria (due to the deficiency of UPG-II cosynthase deficiency), prophyria cutanea tarda (UPG-decarboxylase), hereditary coproporphyria (due to the deficiency of CPG-III oxidase) and hereditary protoporphyria (due to the deficiency of haem synthase). All the porphyrias except AIP have cutaneous photosensitivity urticaria in addition to abdominal pain, nausea and psychiatric manifestations. The cutaneous photosensitivity is due to the accumulation of porphyrins in the skin which as exposure sunlight causes skin scarring. Abdominal pain and neuropsychiatric problems are due to the accumulation of ALA and porphobilinogen in abdominal nerves and brain.

Haem Catabolism

After about 120 days, the RBCs are haemolysed and the haemoglobin comes out of RBCs. Next the globin part of haemoglobin gets separated from haem. The globin is hydrolysed to amino acids and goes to the amino acid pool. The iron of haem is liberated and is reutilised. The remaining part of haem after the removal of iron is the porphyrin.

The further metabolism of porphyrin takes place in the liver, bone marrow and spleen to first form biliverdin and then bilirubin. These changes take place in the reticuloendothelial cells. About 350mg of bilirubin is formed in the body everyday from haem containing proteins like haemoglobin, myoglobin and cytochromes.

The haem is oxidized by haem oxygenase system of microsomes requiring NADPH, oxygen and cytochrome c. The hydrolysis alpha-methenyl

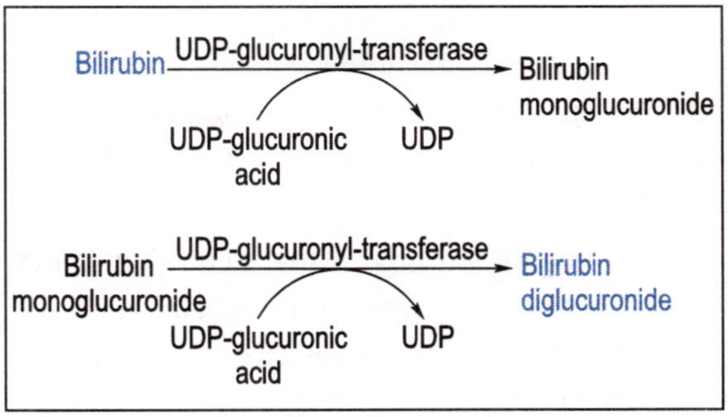

Fig. 7.3: Production of bilirubin diglucuronide

bridges between pyrrole rings I and II forms a linear tetrapyrrole, the green coloured biliviridine. This also leads to the formation of carbon monoxide. The liberated iron is combined with transferrin. The biliverdin is further reduced to the yellow coloured bilirubin by a NADPH dependent reductase.

Next is the binding of plasma albumin with the free bilirubin, for transporting it to the liver where the bilirubin is taken up for further metabolism. The albumin has two binding sites, the low and high affinity binding sites. The free, unconjugated or indirect bilirubin binds to albumin for its transport. As several substances including drugs like aspirin, sulpha drugs, etc. require binding to albumin for their transport, an elevated level of these substances may displace bilirubin from the binding of albumin, and the free bilirubin, thus liberated, may cross the blood brain barrier and cause toxic effects in the brain leading to kernicterus in the new born infants, especially in the premature babies. Hence, care need to be exercised when these drugs have to be given to such newborns.

On reaching the liver the bilirubin, after getting separated from albumin is taken up by the hepatocytes against a concentration gradient. The conjugation of bilirubin with glucuronic acid occurs in the hepatocytes with the help of UDP-glucuronyl transferase which transfers the glucuronyl moiety from UDP-glucuronic acid to bilirubin to generate bilirubin mono glucuronide (BMG). In the next step, one more glucuronyl moiety is transferred, from another mole of UDP-glucuronic acid to bilirubin monoglucuronide, by the same enzyme to form bilirubin diglucuronide (BDG). This process is termed as conjugation of bilirubin.

In the premature infants the UDP-glucuronyl transferase enzyme level is not sufficient and so, the conjugation process is decreased resulting in raised concentrations of free bilirubin which may cause kernicterus. Crigler Najjar syndrome is due to the hereditary absence/decreased level of UDP glucuronyl transferase enzyme, characterised by elevated level of free or unconjugated bilirubin in plasma.

After the conjugation step, the conjugated bilirubin, which is a waste product, is excreted in bile into the duodenum through the bile duct. Hence, it is also known as bile pigment. In the small intestine, the glucuronidase enzyme removes the glucuronic acid from BDG to form bilirubin which is further reduced to urobilinogen. The urobilinogen undergoes enterohepatic cycle and hence, is reabsorbed into the blood and is resecreted into the bile. While in blood, the urobilinogen passes through the kidneys and gets excreted in the urine. Therefore, the normal urine contains minimal quantities of urobilinogen. The excretion urobilinogen is increased in haemolytic jaundice, it is normal or decreased in hepatic jaundice and is absent in obstructive jaundice, as the bile flow to the duodenum is obstructed. A major share of urobilinogen is converted to stercobilinogen and gets excreted in the faeces giving it a specific colour. In the obstructive jaundice, the stool becomes clay coloured due to the lack of stercobilinogen in it.

Jaundice

The normal serum total bilirubin level is less than 1mg/dl out of which the direct (conjugated) bilirubin is about 0.2 to 0.4mg and the indirect (unconjugated or free) bilirubin is 0.4 to 0.6mg/dl. Hyperbilirubinaemia or jaundice results when the serum total bilirubin is more than 1mg/dl. One way to classify hyperbilirubinaemia is unconjugated hyperbilirubinaemia and conjugated hyperbilirubinaemia and hyperbilirubinaemia in which both conjugated and unconjugated are elevated.

A better way to classify jaundice is (a) haemolytic, (b) hepatic and (c) obstructive, or, otherwise called as prehepatic, hepatic and post-hepatic respectively. The jaundice can be either congenital or acquired.

Hemolytic jaundice

This results due to the excessive haemolysis of erythrocytes due to several causes like haemoglobinopathies, thalassaemias, erythroblastosis fetalis, any other abnormalities in RBCs or due to enzyme deficiencies like glucose-6-phosphate dehydrogenase. This leads to the elevated levels of bilirubin in the serum, which the normal liver is not able to take up and conjugate with glucuronic acid for the purpose of excretion. This way, the free (unconjugated/indirect) bilirubin is increased in the serum. The van den Bergh's test will be indirect positive in haemolytic jaundice. When the diazotised sulphanilic acid is reacted with the patient's serum containing free bilirubin no purple colour is obtained because free bilirubin is not soluble in water, but, when methanol is added to the reaction mixture, so as to make the free bilirubin water soluble, a purple colour is developed. So, in haemolytic jaundice, the free bilirubin level is increased, because the liver, although functioning normally, is unable to process the excess of bilirubin presented to it. The serum enzymes like the transaminases and alkaline phosphatase are within normal limits in haemolytic jaundice.

Obstructive jaundice

The obstructive jaundice is due to the obstruction

to the bile flow may be as a result of a tumour. Here the uptake of bilirubin and its conjugation by the liver is normal but the excretion of conjugated bilirubin is defective as a result of obstruction to the bile flow. Therefore, the conjugated bilirubin regurgitates into the blood. The van den Bergh's test would be direct positive here because a purple colour is developed on reaction between the serum of these patients and diazotised sulphanilic acid, as the direct bilirubin is water soluble. Therefore, the conjugated bilirubin is elevated in patients with obstructive jaundice. The urobilinogen is not formed in such patients and hence, urine urobilinogen test will be negative in them. The stool will be clay coloured due to the absence of stercobilinogen. Out of the serum enzymes, the alkaline phosphatase is elevated, whereas the serum transaminases (ALT & AST) are normal in concentration.

Hepatic jaundice

Hepatic jaundice or hepatocellular jaundice occurs due to hepatitis viral infection or due to cirrhosis of liver. Here initially the free bilirubin is increased as the damaged liver is unable to take up and conjugate the bilirubin. Later on the inflammation develops in the hepatocytes and biliary canaliculi causes a component of obstruction. Due to this obstruction the conjugated bilirubin also increases. Therefore, the van den Bergh's test would be biphasic, meaning, both free and conjugated bilirubin are elevated. When the serum of such a patient is mixed with the reagent, a purple colour develops directly and the colour intensifies on addition of methanol showing that both direct and indirect (conjugated and free) bilirubins are present. This is known as biphasic reaction. Further tests for diagnosing hepatocellular jaundice include serum transaminases ALT, AST and ALP which are all increased in hepatitis.

chapter 8

HAEMOGLOBINOPATHIES AND THALASSAEMIAS

Several abnormal haemoglobins have been identified, the most common of which is sickle cell haemoglobin (Hb S). The haemoglobinopathies are due to one base substitutions, like transition or transversion. This results in an altered sequence of amino acids, only one amino acid being replaced by another. Thalassaemias are due to disproportionate synthesis of alpha or beta chains. In beta thalassaemia, the beta globin chains are formed less or they are completely absent and the alpha thalassaemia is one which has less or absent synthesis of alpha chains. Stated differently, thalassaemia is caused by decreased formation of either alpha or beta chains, but with a normal sequence.

Haemoglobinopathies

Sickle cell haemoglobin is the result of the substitution of an adenine by thymine in the 6th codon of beta chain gene. Because of this, the 6th amino acid (glutamic acid) in the beta globin chain is substituted by valine. The acidic amino acid is replaced by a neutral amino acid due to which a "sticky patch" is developed on the Hb S molecule. This sticky patch on one Hb S attaches with a corresponding patch on another Hb S, and this starts polymerisation of Hb S molecules which subsequently get precipitated in the RBC, thereby changing the shape of the erythrocytes to sickle shape which can be visualised under a microscope.

The sickle cell RBCs, while passing through the capillaries, get haemolysed and this leads to haemolytic jaundice. The Hb S abnormally binds and transports oxygen. Due to this there is hypoxia in the tissues and that leads to chronic headache and body pain. Recurrent infections are also common in sickle cell disease. As the sickle RBCs can form thrombus, such patients may suffer form stroke.

The sickling is more prevalent in the deoxygenated state. In Hb S trait 50 per cent of Hb is normal (Hb A) and the other 50 per cent is Hb S. This is a heterozygous state, and the patient can lead near to a normal life, but such people are carriers of the Hb S gene and cause the abnormality in the offsprings, provided this person marries one having similar type of abnormality. In a marriage between Hb S trait male and female, their children will be 25 per cent normal, 25 per cent will be homozygous with Hb S disease and the remaining 50 per cent will be heterozygous with Hb S trait. The Hb S disease is a homozygous condition in which all the haemoglobin molecules are of Hb S type.

The patients will be needing frequent blood transfusions which could be causing iron overload and haemochromatosis and haemosiderosis, leading to liver cirrhosis, which may have to be treated by phlebotomy. Keeping haemoglobin in the oxygenated state reduces sickling and could be useful. Administration of thiourea also reduces sickling. The infection requires treatment with antibiotics. Body pain can be relieved by analgesics.

Other types of haemoglobinopathies include Hb C (replacement of 6th glutamic acid in the beta chain by lysine). Hb E (replacement of 26th glutamate in beta chain by lysine) and Hb D (replacement 121st glutamic acid by glutamine. The homozygous Hb C (CC) and double heterozygotes for Hb S and Hb C (SC) are moderately to severely anaemic.

Thalassaemias

As the haemoglobinopathies result from the changes in the sequence of amino acids in the beta chain of haemoglobin, thalassaemias are caused by decreased production of either alpha or beta chains. Here the structure of haemoglobin is normal.

In alpha thalassaemia, the synthesis of alpha globin chain is decreased or absent due to a deletional mutation or due to a defective regulator gene. The complete absence of alpha chaina results in alpha thalassaemia Hb-Bart disease and 50 per cent presence of alpha chain causes alpha thalassaemia trait. There are 4 alleles of alpha globin genes and if one or two genes are deficient, alpha thalassaemia trait results and if 3 or 4 alleles are defective alpha thalassaemia Hb-Bart (β_4) disease could result further, in foetal death, as the foetus is not able to synthesise Hb F due to the lack of alpha globin chains. This condition is called as *hydrops foetalis*.

Beta thalassaemia develops when the beta chain synthesis is decreased. There are two alleles for the beta globin gene. If one of the alleles is deficient, beta thalassaemia minor is caused. When both the alleles are deficient, beta thalassaemia major is resulted. A newborn infant with beta thalassaemia major will look normal at birth, but will start developing anaemia within one year and will have to be given blood transfusion which can lead to iron overload. The Hb A concentration in the newborn infant is 15 to 20 per cent and hence, there is less requirement of beta globin chains in the foetal life. In the first few weeks of life the Hb F has to be replaced by Hb A, containing beta globin chains in place of gamma globin chains.

chapter 9

FAT SOLUBLE VITAMINS

The vitamins are micronutrients present in food items. They cannot be synthesised in the body but are required in very small amounts, the deficiency of which causes different pathological states. All the vitamins are present in an average Indian diet and so there is very little need for any extra supplements provided one is taking a balanced diet consisting of enough fruits and vegetables.

There are two groups of vitamins, fat soluble and water soluble varieties.

The fat soluble vitamins are A, D, E and K.

VITAMIN A

The vitamin A is present only in the animals. The plants have the precursor of vitamin A known as *beta carotene*. Beta carotene is made up of 2 beta ionone rings attached with isoprenoid units. One molecule of beta carotene can be cleaved at the centre to form two vitamin A molecules.

The active forms of vitamin A are retinol (the alcohol form), retinaldehyde (the aldehyde form) and retinoic acid (the acid form). It is possible to inter-convert retinol and retinal but retinoic acid, when formed from retinal, cannot be converted back to the other forms. The retinol reductase oxidises retinol to retinal and back.

Sources of vitamin A

The vegetarian sources of beta carotene are the yellow vegetables and fruits like tomatoes, papaya, pumpkins, mangoes, green leafy vegetables and carrots. The vitamin A is present in fish liver oils and milk and milk products.

Dietary allowance

The daily dietary requirement of vitamin A in children is about 550 microgram, women about 700 microgram, pregnant women about 1000 microgram and men also need about the same quantity.

Digestion and absorption

The beta carotene in the intestine is cleaved to retinal molecules by a dioxygenase. The retinol formed from retinal is absorbed with other lipids into the intestinal mucosal cells through micelle formation, with the help of bile salts, in the absence of which vitamin absorption is affected. In the intestinal mucosal cell, the vitamin A along with other lipids, are packaged into chylomicrons which passes into the lymphatics and then to blood. Liver is the storage location of vitamin A.

Functions of vitamin A

The three active forms of vitamin A, retinol, retinal and retinoic acid have different functions in different tissues.

The retinol is transported in the plasma by binding to retinol binding protein to various tissues like retina and other target tissues. Inside the cells retinol is transported by cellular retinol binding protein.

In the gonads the retinol has a reproductive role and acts like a steroid hormone. In the retina the retinal is crucial in the visual process. The retinoic acid in the skin and epithelial tissue helps to maintain a healthy skin and a healthy and moist epithelium and mucosal membrane. The retinol has to bind to the hormone response elements of the genes to start transcription. The beta carotene and vitamin A have antioxidant properties and are also proved to have anticancer effect as the vitamin A is required for a healthy epithelium. The beta-carotene is also helpful in preventing CHD.

Wald's visual cycle

The rhodopsin is made up of the protein, opsin and 11-cis-retinal. It is in the rod cells of the retina that rhodopsin is located. The light falling on retina converts cis-retinal to trans-retinal. The photons falling on the retina leads to conformational changes of the rhodopsin which ultimately forms opsin and trans-retinal through a series of intermediates. The nerve impulse is further transmitted to the brain for the process of vision. The trans retinal is converted back to cis-retinal in the dim light and further attaches to opsin and regenerates the visual pigment rhodopsin.

The bright light can exhaust the rhodopsin of the rod cells and hence, a person entering a dark room or when the current goes away in the night, people take some time to see the things around and this time is called as "the dark adaptation time". During this period, lasting about a few minutes, the rhodopsin stores are replenished and resultantly the vision becomes better. In people with a deficiency of vitamin A the dark adaptation time is increased proportionately. Hence, it is recommended that the drivers driving at night should undergo dark adaptation time testing, and if need be vitamin A supplements can be given.

Rods are for vision in dim light and cones are involved in the colour vision.

Deficiency of vitamin A

Nyctalopia or night blindness resulting from lengthened dark light adaptation is due to deficiency of vitamin A. In xerophthalmia, another deficiency manifestation of vitamin A, the conjunctiva and cornea get keratinised, gradually becoming opaque, leading to keratomalacia. The formation of bitot's spots can be seen as thickenings on the cornea. There are a number of children with vitamin A deficiency. It is also called as anti-infection vitamin as its lack may cause infection by micro-organisms.

If taken in large quantities, hypervitaminosis can be resulted, having symptoms like irritability, headache and nausea.

VITAMIN D

This vitamin can be formed from 7-dehydrocholesterol and ergosterol is derived from plant in presence of UV light. The 7-dehydrocholesterol in the skin is converted to cholecalciferol in the presence of uv light. The cholecalciferol is transported to the liver where hydroxylation at 25th position occurs to synthesise 25-hydroxycholecalciferol in presence of the enzyme 25-hydroxylase. Next the 25-hydroxy cholecalciferol is transported to the kidneys where 1-alpha-hydroxylase enzyme, which is activated by parathyroid hormone, hydroxylates 25-hydroxy cholecalciferol to 1, 25-dihydroxycholecalciferol which is the active form of vitamin D, being called as *calcitriol*.

The calcitriol has several functions and its target tissues are bones, small intestine and kidneys. In the bones, calcitriol causes the mineralisation of bones by stimulating the osteoblastic activity. The calcitriol increases the serum calcium levels. In the intestinal mucosal cell, the calcitriol brings about the synthesis of calcium binding protein (calbindin) which binds with the calcium in the intestinal lumen and helps in the absorption of cal-

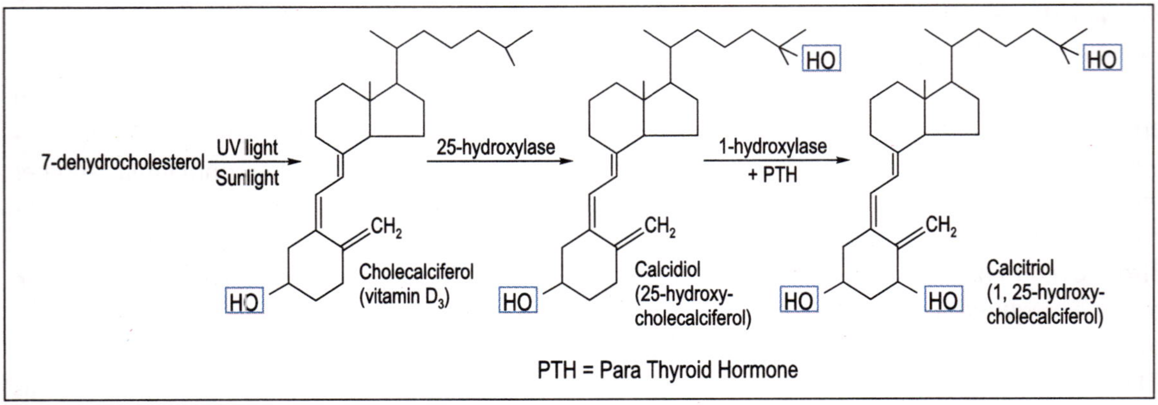

Fig. 9.1: Synthesis of Calcitriol

cium. The function of calcitriol in the kidneys is to reabsorb calcium from the glomerular filtrate. Due to these actions, calcitriol ultimately leads to an increase of serum calcium.

The normal serum calcium level is 9-11mg/dl, out of which about 5mg/dl is in the ionised state which is the active form of calcium. A decrease in the level of ionised calcium causes tetany. The remainder of the calcium is bound to proteins.

The parathyroid hormone causes demineralisation of the bones, whereas calcitonin is responsible for mineralisation of bone. The calcitriol, parathyroid hormone and calcitonin function to maintain the calcium homeostasis, because the former two increases serum calcium level, and calcitonin decreases the serum calcium level.

Deficiency of vitamin D

Rickets in children is due to the deficiency of vitamin D. The features of rickets include soft and pliable bones which may be fractured easily. There may be knock knees, bow legs and pigeon chest. The bone alkaline phosphatase isoenzyme in serum is found to be raised and the serum calcium and phosphorous level are decreased.

In the adults, the vitamin D deficiency causes osteomalacia. Here also the calcium and phosphate show slight decrease with elevated concentrations of alkaline phosphatase.

The vitamin D deficiency can be due to lack of exposure to sunlight.

Hypervitaminosis of vitamin D can be produced because of high intake of the vitamin D.

VITAMIN E

The alpha tocopherol is known as vitamin E. The function of tocopherol is to counter the effect of free radicals which are highly reactive and can degrade a variety of biomolecules. The free radicals are produced during various metabolic reactions which, if not neutralised, can cause deleterious effects on various molecules like proteins, membranes, DNA, etc. The free radicals are also implicated in the ageing process. The selenium which is associated with an antioxidant enzyme, glutathione peroxidise, is found to act along with vitamin E.

The dietary allowance of vitamin E is about 8-12mg/day. We have to take more vitamin E when the diet has more unsaturated fatty acids so as to prevent lipid peroxidation and free radical production.

Vegetable oils are good sources of vitamin E.

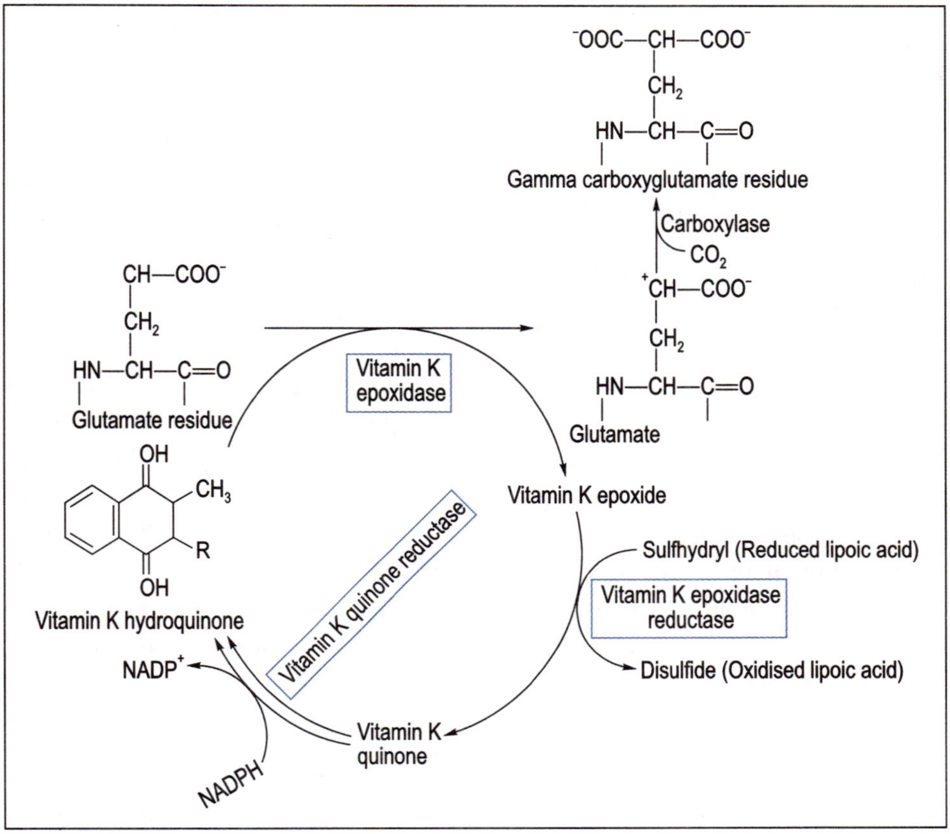

Fig. 9.2: The role of vitamin K in synthesis of gamma-carboxyglutamate

VITAMIN K

Vitamin K is essential for blood coagulation. It is required for the activation of coagulation factors II, VII, IX and X. When these factors are formed, these are inactive and have to be activated when a requirement arises on an emergency basis. This is an example of post-translational modification and requires vitamin K. This also prevents unnecessary clot formation.

Vitamin K is involved in the carboxylation of glutamic acid residue at its gamma carbon atom so that the glutamate now possesses 2 negatively charged carboxylic acid groups, to which the positively charged divalent calcium can be attached. This way the clotting factors are activated.

Several other tissue proteins also need vitamin K dependent gamma carboxylation for their activation. One such example is a protein of bones, osteocalcin, which needs gamma carboxylation mediated by vitamin K for its activation.

Defective digestion and absorption of lipids can cause the deficiency of all the fat soluble vitamins including vitamin K.

Warfarin and dicoumarol act as competitive inhibitors of the gamma carboxylation of glutamic residues of the clotting factors and hence, inhibit the coagulation process.

chapter 10

WATER SOLUBLE VITAMINS

All the water soluble vitamins function as coenzymes in various metabolic reactions. They are thiamine, riboflavin, niacin, pyridoxine, pantothenic acid, biotin, ascorbic acid, folic acid and vitamin B_{12}.

THIAMINE

The husk of cereals like wheat and manually pounded rice have higher quantities of vitamin B_1 as the husk is rich in that vitamin. Hence, the nutritive quality of the whole cereals is far better than the refined foods. The coenzyme form of thiamine is thiamine pyrophosphate (TPP) and consists of a pyrimidine attached to a thiazole ring and two phosphates.

The TPP serves as the coenzyme for (1) pyruvate dehydrogenase which oxidatively decarboxylates pyruvate to acetyl CoA and (2) alpha-keto glutarate dehydrogenase which oxidatively decarboxylate alpha-keto-glutarate to succinyl CoA.

The two transketolase enzymes in the hexose monophosphate shunt pathway also require TPP.

The deficiency states of thiamine are beriberi and polyneuritis. The symptoms of beriberi are fatigue and anorexia. In wet beriberi oedema and cardiovascular components are present, whereas in dry beriberi central nervous system disturbances are predominant with peripheral neuritis. Deficiency of thiamine causes lactacidosis.

It can be noticed that thiamine requirement is for the metabolism of carbohydrates.

RIBOFLAVIN

The coenzymes of riboflavin are flavin mononucleotide (FMN) and flavin adenine dinucleotide (FAD). FAD functions in the dehydrogenation reactions, for example, in the dehydrogenation of succinate to fumarate while the FAD becomes $FADH_2$. The FAD and FMN are involved in the respiratory chain also. The FMN is required by amino acid oxidases.

The symptoms of riboflavin deficiency are angular stomatitis and glossitis.

NIACIN

The nicotinamide adenine dinucleotide (NAD) and nicotinamide adenine dinucleotide phosphate (NADP) are the coenzyme forms of niacin. The NAD is used for dehydrogenation reaction, for example, lactate dehydrogenase which converts lactate to pyruvate, while NAD becomes NADH + H^+. Several other dehydrogenase enzymes also utilise NAD, for example, hydroxy acyl DH. An NADH mole, if oxidised in the electron transport chain, can synthesise 2.5 ATP molecules.

The enzymes that produce NADPH are in the HMP shunt pathway. These enzymes are glucose-6 phosphate dehydrogenase and 6-phosphogluconate dehydrogenase. The other two minor contributions of NADPH are made by malic enzyme and isocitrate dehydrogenase, present in cytoplasm. It may be remembered that the main isocitrate dehydrogenase is located in the mitochondria functioning in the TCA cycle.

The NADPH has several important functions like maintenance of the integrity of RBC membrane, reducing oxidised glutathione to reduced glutathione, conversion of methaemoglobin to haemoglobin, synthesis of cholesterol, (for example HMG CoA reductase) steroid hormones, bile acids and fatty acid synthesis.

The deficiency of niacin leads to pellagra which manifests in dermatitis, diarrhoea and dementia.

The niacin can be formed from tryptophan. Hence, people taking a food lacking in tryptophan (a diet based on maize) may suffer from pellagra.

Nicotinic acid decreases lipolysis in adipose tissue and hence, indirectly decreases the synthesis of cholesterol from acetyl CoA which is produced from the beta oxidation of fatty acids.

PYRIDOXINE

The pyridoxine is known as vitamin B_6. The pyridoxal phosphate is the active form of pyridoxine and is required for the metabolism of amino acids. The pyridoxal phosphate is the coenzyme for transamination and decarboxylation of amino acids. It is involved in the conversion of tryptophan to niacin, in the synthesis of haem and in glycogenolysis.

The transaminases, aspartate aminotransferase (AST/GOT) and alanine aminotransferase (ALT/GPT) require pyridoxal phosphate. The AST catalyses the transamination of amino group between alpha ketoglutarate and aspartic acid to generate glutamic acid and oxaloacetate. The ALT is required for the transfer of amino group between alpha keto glutarate and alanine so as to form glutamic acid and pyruvic acid.

Amino acid decarboxylases also have the need of pyridoxal phosphate (PLP). The examples are the formation of inhibitory neurotransmitter, gamma amino butyric acid form glutamate by decarboxylation. The synthesis of histamine, a mediator of allergic responses, from histidine and the formation of serotonin (the relaxing hormone) from 5-hydroxy tryptophan are examples.

As mentioned, the initial reactions of haem synthesis require PLP. Further the deficiency of PLP can lead to homocysteine excess which is a risk factor of CAD. The PLP is necessary for the metabolism of sulphur containing amino acids as both cystathionine synthesis and further metabolism requires PLP.

PLP is essential for the action of the rate limiting enzyme of glycogenolysis, the glycogen phosphorylase.

Deficiency symptoms

The pyridoxine deficiency may lead to neurological symptoms as it will adversely affect the synthesis of several neurotransmitters like GABA, dopamine, serotonin, etc. Peripheral neuritis is seen in PLP deficiency as it can result in defective synthesis of sphingolipids which culminates in demyelination. PLP lack can create pellagra like symptoms.

PANTOTHENIC ACID

The pantothenic acid is prepared with pantoic acid and beta-alanine. The coenzyme A contains pantothenic acid, beta mercapto ethanolamine and AMP. There are several CoA derivatives like acetyl CoA, acyl CoA, HMG-CoA and succinyl CoA. The acyl carrier protein (ACP) of fatty acid synthase complex also has pantothenic acid.

The deficiency symptoms of pantothenic acid in humans are rare.

BIOTIN

The biotin is used as a coenzyme in carboxylation reactions. The examples include pyruvate carboxylase, which adds CO_2 to pyruvate to form oxaloacetate, in presence of ATP which becomes

ADP and Pi. The biotin first combines with CO_2 to form carboxy- biotin, which then passes the CO_2 to pyruvate to form oxaloacetate. This is in the gluconeogenic pathway. Other reactions involving biotin are acetyl CoA carboxylase which converts acetyl CoA to malonyl CoA, a step in the fatty acid synthesis. Another reaction is propionyl CoA carboxylase to convert propionyl CoA, resulting from the beta oxidation of odd number chain fatty acids, to methyl malonyl CoA. All these reactions require ATP and CO_2 in addition to biotin.

Eating raw eggs is not advisable because it contains avidin protein which can bind to biotin and cause the deficiency of biotin. As the avidin-biotin binding is quite strong, this is used for techniques like enzyme-linked immunoassays. Streptavidin also can bind with biotin and can also find use in assays of different constituents in body fluids.

FOLIC ACID

The folic acid contains pteroyl glutamic acid linked to para-aminobenzoic acid. Folic acid is transported to liver, after its absorption.

The folic acid is first reduced to dihydro folate and then to tetrahydrofolate by folate reductase, the hydrogen for reduction is contributed by NADPH. The function of tetrahydro folate (THFA) is to take part in the 1 carbon metabolism. There are several one carbon residues carried by THFA like methenyl, methelene, methyl, hydroxymethyl, formimino and formyl. It is possible to interconvert all the one-carbon units except methyl THFA which, when formed, cannot be converted to the other forms. The methyl group can be accepted by only vitamin B_{12}. Hence, vitamin B_{12} deficiency leads to methyl folate trap and, therefore, causes accompanying folate deficiency also. The methyl cobalamine gives the methyl group to homocysteine to form metionine. The methyl group of metionine is used to synthesise choline, epinephrine, etc.

The synthesis of purine nucleotides require methenyl THFA and formyl THFA. Hence, the synthesis of nucleic acids are affected in folate deficiency. The synthesis of haemoglobin is unaffected. This leads to immature nucleus in RBCs. This is known as macrocytic anaemia.

The leafy vegetables provide good amounts of folate which is absorbed from the jejunum. The pregnant women should take enough leafy vegetables to prevent birth defects in foetus. Folic acid and vitamin B_{12} need to be given to take care of vitamin B_{12} deficiency induced folate deficiency.

VITAMIN B_{12}

Previously vitamin B_{12} used to be known as extrinsic factor but now it is more correctly called as cobalamin. It has acorrin ring structure with pyrrole rings. There is a cobalt atom in the ring. The vitamin B_{12} has benzimadozole, cyanide, methyl,

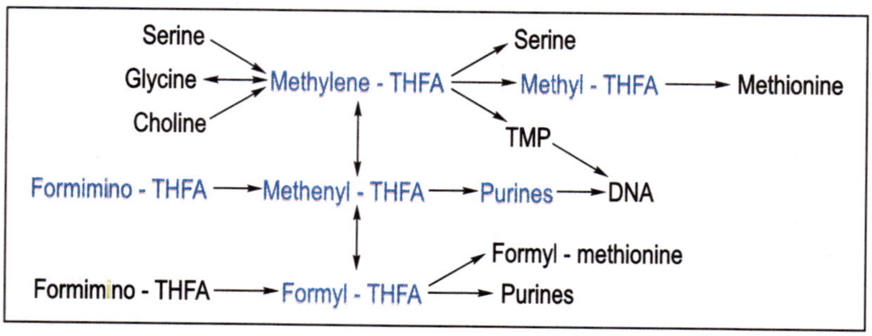

Fig. 10.1: One-carbon units

adenosyl or hydroxyl group in it so that respectively we have cobalamin, cyanocobalamin, methylcobalamin, adenosylcobalamin and hydroxycobalamin. The methylcobalamin and adenosylcobalamin function as vitamin B_{12} coenzymes.

The absorption of vitamin B_{12} taking place in the ileum requires binding with intrinsic factor—a protein factor. The transport of vitamin B_{12} in the blood is by combining with a glycoprotein, the transcobalamin. None of the water soluble vitamins are stored in the body, but a few milligram quantities of adenosylcobalamin can be stored in the liver which could last for couple of years.

Coenzyme role of vitamin B_{12}

There are two enzymes which need vitamin B_{12} for their action. One is in the conversion of L-methyl malonyl CoA to succinyl CoA catalysed by the enzyme methyl malonyl CoA mutase. This enzyme is in the metabolism of propionyl CoA derived from the metabolism of odd chain fatty acids and from branched chain amino acids and methionine.

Another vitamin B_{12} requiring enzyme is in the formation of methionine from homocysteine by transferring the methyl group from methyl cobalamine and the enzyme is a methyl transferase.

As mentioned in the section on folic acid, the deficiency of vitamin B_{12} can cause a deficiency of folate also because the methyl-THFA cannot be converted to other one-carbon forms, and the sole way of freeing folate from methyl-THFA is by vitamin B_{12}. Hence, vitamin B_{12} deficiency leads to methyl folate trap and subsequent folate deficiency.

Deficiency of Vitamin B_{12}

It is quite common in vegetarian Indians who do not consume enough milk or milk products, the vitamin being absent in the plant foods. The absence of intrinsic factor, as in gastrectomy or gastric atrophy can lead to impaired absorption of vitamin B_{12}. As the requirement increases in pregnancy and lactation, the folate deficiency occurs.

Deficiency symptoms

Folate deficiency produces macrocytic anaemia, whereas vitamin B_{12} deficiency causes megaloblastic anaemia. Due to the deficiency of vitamin B_{12} methyl malonic aciduria occurs and methyl malonic acid can go to the brain and cause demyelination leading to neurological manifestations. Further the deficiency of methionine synthesis also can hamper myelination due to the lack of trans lipids. The demyelination can affect cerebral cortex and spinal cord and hence, is called as combined degeneration.

ASCORBIC ACID

Ascorbic acid, also called as vitamin C, can be synthesised by several animals and plants too from glucose, but humans cannot synthesise vitamin C due to want of the enzyme gulonolactone oxidase. So, the ascorbic acid has to be taken in the diet. The good sources include lemons, gooseberry, leafy vegetables and fruits.

Metabolic functions

The vitamin C is involved in the post-translational modification of procollagen to collagen. It is required for keeping iron in the ferrous state, as ferrous iron is essential for the enzymes lysyl hydroxylase and prolyl hydroxylase which hydroxylates lysine to hydroxy lysine, and proline to hydroxyl proline respectively. This is crucial for the maturation of collagen especially for the cross-linking between the polypeptide chains.

As the vitamin C reduces the ferric iron to ferrous, it helps in the absorption of iron from the GI tract. It converts methaemoglobin to haemoglobin. Vitamin C takes part in the formation of serotonin, and is necessary for the metabolism of aromatic amino acids, synthesis of bile acids and for preventing cataract development.

Ascorbic acid, along with tocopherol, beta carotene and vitamin A are potent antioxidant vitamins which serve to neutralise the effect of free radicals having deleterious effects including cancer. The other antioxidants comprises of glutathione, glutathione peroxidase, glutathione reductase, selenium and superoxide dismutase.

The vitamin C has wound healing properties.

Deficiency states of vitamin C

Severe deficiency of vitamin D causes bleeding of gums and scurvy. This is due to the immature collagen, in the walls of the capillaries which is the result of the deficiency of ascorbic acid. This can also cause internal haemorrhage.

Bones can be deformed and weak. This is again due to defective hydroxylation of lysine and proline of collagen leading to lack of cross-linking so that the bones become brittle and pliable.

Daily requirement

The daily requirement vitamin C is about 75-100 mg/day with additional requirement in pregnancy and lactation.

chapter 11

MINERAL METABOLISM

The minerals are important to maintain proper health. The macro-elements are required in quantities to the tune of more than 50mg or more per day.

CALCIUM

Out of all the minerals, calcium is the one that is present in highest quantities in the body. The major portion of body calcium is present in the bones and teeth, with very little present in body fluids and cells. The total amount of calcium in the body is about 1300 grams.

The children should take about 1500mg of calcium, adults about 1000mg and during pregnancy and lactation, the intake should be hiked to 1600mg per day.

The absorption of calcium requires calbindin protein which is under the influence of calcitriol, the active form of vitamin D. The parathyroid hormone (PTH) activates vitamin D, by stimulating alpha-1 hydroxylase which forms calcitriol.

The dietary substances decreasing calcium absorption are phytates, phosphates and oxalates.

Functions of calcium

The normal serum calcium level is 9-11mg/dl. The ionised calcium, the diffusible form, which is also the metabolically active form is about 4.8-5.2mg/dl (about 60 per cent of total plasma calcium). The influence of parathyroid hormone is fully on the ionised calcium levels. A decrease in ionised calcium resulting from hypoparathyroidism can precipitate increased irritability of the peripheral nerves, initiating muscle spasms leading to generalised convulsions culminating in tetany.

The remaining (about 40 per cent) plasma calcium is bound to proteins especially to plasma albumin, representing the nondiffusible calcium. The concentration of nondiffusible calcium can change depending on the plasma albumin levels. In hypoalbuminaemia the total plasma calcium is less than the normal, still tetany may not occur, as the ionised calcium levels are unaffected.

The bones contain deposits of calcium phosphate in a collagen matrix. A major part of calcium is in the bones. Bones also supplies calcium for maintaining serum calcium level between 9-11mg/dl. Calcitriol and parathyroid hormone (PTH), secreted by parathyroid glands of the thyroid, increase serum calcium levels. The calcitonin, secreted by thyroid gland decreases calcium level in serum, but helps in the deposition of calcium in the bones. The PTH causes demineralisation of bones. Calcitriol also increases the absorption of calcium from the GI tract, by increasing the synthesis of calbindin protein, and the reabsorption of calcium from glomerular filtrate.

Muscle contraction is activated by calcium. On nerve stimulus, sarcoplasmic reticulum liberates calcium which interacts with troponin C and this initiates muscle contraction by activating actin and myosin.

Transmission of nerve impulse across synaptosomes is dependent on calcium. The secretion of various hormones also need calcium.

Calcium combines with the two carboxyl groups of gamma carboxyglutamic acid residues (formed in the presence of vitamin K) of clotting factors including prothrombin to activate the coagulation process. Several other enzymes are also activated by calcium. For example, calcium calmodulin complex activates kinases which in turn activate other enzymes like phosphorylase kinase to bring about specific metabolic responses, for example, glycogenolysis.

The role of calcitriol in the metabolism of calcium has already been explained in the section of vitamin D.

PTH and CALCITONIN

The pro-PTH when synthesised is a large molecule of 84 amino acids which has to undergo post translational changes to form the active PTH in a way analogous with the activation of proinsulin to insulin. When the serum calcium level falls below normal the PTH is stimulated and this leads to the demineralisation of bones, thereby, increasing serum calcium levels. As mentioned earlier, the PTH activates the conversion of 25-hydroxy cholecalciferol to 1, 25-dihydroxy cholecalciferol (calcitriol) which increases serum calcium level by two ways: absorption of calcium from intestine and reabsorption of calcium from glomerular filtrate. When the calcium level rises, the secretion of PTH decreases. Thus, the bone is subjected to recalcification and decalcification cycles, which is the result of various factors like PTH, calcitonin, calcitriol, vitamins A, C and B_{12}.

The calcitonin is a hormone secreted by thyroid tissue and this helps in the bone mineralisation and so it decreases serum calcium level.

Rickets and osteomalacia

These topics have already been covered under vitamin D deficiency diseases.

IRON METABOLISM

Iron is an important metal present in haemoglobin, myoglobin and cytochromes and is required in certain enzyme reactions. The women have to consume more iron during their reproductive period and accordingly the daily requirement of iron in them is about 20mg. The males have to ingest about 10mg of iron everyday.

Sources of iron

Leafy vegetables, nuts, wheat, meat, eggs and fishes are good sources of iron.

Absorption of iron

Only about 10-15 per cent of the ingested iron is absorbed because, when once absorbed only very little of iron can be excreted and hence, to prevent the toxic accumulation of iron in the body tissues, it is pertinent to control the iron content of the body at the level of absorption itself. Hence, the iron is sometime referred to as a "one way metal". Therefore, only about 2-3mg of iron are absorbed daily.

The gastric acids release iron from complexes. The iron is reduced from ferric to ferrous by reducing agents like vitamin C. The ferrous iron is absorbed from the stomach and duodenum. The ferrous iron absorbed into the mucosal cells, gets oxidised to ferric iron and combines with apoferritin to form ferritin and is stored in the mucosal cell. A portion of this iron present in ferritin is lost when the cells shed into the intestinal lumen. The way by which the iron absorption is regulated is called as mucosal block theory. The limited availability of apoferritin molecules in the intestinal mucosal cells helps to curtail iron absorption.

The stored ferric iron from ferritin is released and reduced to the ferrous form and passes into the plasma to complex with transferrin protein, after further oxidation to ferric state. The transferrin carrying ferric iron is taken up by bone marrow to synthesise haemoglobin. The transferrin transports iron to all the tissues requiring iron for various functions.

Normal plasma level of transferrin is about 240-260mg/dl which increases in iron deficiency anaemia. The total iron binding capacity (TIBC) is about 380mg and one-third of this capacity is saturated with iron. This is the protein bound iron which is about 130mg/dl. In iron deficiency anaemia TIBC is increased but the serum iron is decreased.

Haemochromatosis is a condition in which iron gets deposited in tissues like skin, pancreas and liver. Increased absorption of iron over a period of time can lead to haemosiderosis and haemochromatosis.

Very little iron is lost in males. Due to the menstrual loss of blood, females lose more iron and hence, the need to take more iron by females.

METABOLISM OF ELECTROLYTES AND WATER BALANCE

Sodium, potassium, chloride and bicarbonates are the major electrolytes of the extracellular and intra cellular fluids. The concentrations of sodium, potassium chloride and bicarbonates are 142meq/L, 5meq/L, 103meq/L and 27meq/L respectively. The potassium is the major cation of intracellular fluid (ICF), while sodium is the major cation of extracellular fluid (ECF). They regulate the pH of the body fluids, take part in the chloride shift, maintain the osmotic pressure and regulate the different compartments of body water in addition to helping the neuromuscular functions. All these electrolytes have to be maintained within a narrow range. The normal serum sodium concentration is 135-145meq/L, normal range of potassium is 3.5-5meq/L, that of chloride is 96-106meq/L and that of bicarbonate is 24-27 meq/L.

Regulation of water balance

The aldosterone secreted by the adrenal cortex causes sodium retention. The hypothalamus secretes the anti-diuretic hormone (ADH) in response to an increase of plasma osmolality. Due to this water reabsorption occurs in the renal tubules resulting in the retention of water and sodium. The ADH secretion will be stopped with a decrease in the osmolality.

The rennin angiotensin system also plays an important role in the water balance in the body.

SODIUM METABOLISM

Sodium regulates the extracellular fluid volume and is very much involved in the water balance. Sodium-potassium ATPase pump functions to keep sodium in the extracellular compartment of the body. A large quantity of sodium (780-800 grams) is filtered at the glomerules, but most of it is reabsorbed in the proximal convoluted tubules. Along with sodium water is also reabsorbed. As mentioned earlier, aldosterone increases sodium reabsorption from distal tubules, while ADH is responsible for the reabsorption of water.

The atrial natriuretic peptides secreted due to atrial stretching removes sodium in urine. In congestive cardiac failure, sodium and water content of the body increases and this leads to oedema.

Hypernatraemia

Elevated level of sodium, called as hypernatremia, though not common, is characterised by symptoms like fever and thirst. It can be encountered in long term cortisone therapy, Cushing's disease and pregnancy.

Hyponatraemia

This is a condition in which the serum sodium level is reduced. Two types of hyponatraemia can be there, one with absolute loss of sodium and second due to excessive hydration. The first type can be because of vomiting and diarrhoea, renal loss of sodium and Addison's disease.

POTASSIUM METABOLISM

Potassium ions are necessary for the working of cardiac and skeletal muscles, the potassium moves out of muscle cells into the ECF during contraction of muscles and the potassium returns to the muscle cells after the muscle contraction. It is essential for nerve function. Potassium is involved in glycogen storage. Low potassium levels can severely affect cardiac muscle activity and can cause muscle weakness.

Hyperkalaemia

The serum potassium levels are elevated in hyperkalaemia and may be due to renal failure, dehydration, Addison's disease and bradycardia.

Hypokalaemia

This is decreased level of serum potassium. The long term hypokalaemia can damage myocardium and kidneys.

Hypokalaemia can result from excessive vomiting and diarrhoea, Cushing's syndrome, heart failure, diuretic therapy and also during the management of diabetic coma with insulin.

OTHER MINERALS OF IMPORTANCE

The various other mineral compounds like magnesium, copper, manganese (functioning as co-enzymes in various enzyme reactions) zinc (in insulin function), molybdenum (component of xanthine oxidase of purine catabolism), iodine (for the synthesis of the thyroid hormones) and selenium (for selenocysteine formation) and also to function as an antioxidant along with vitamin E are of great importance to the body functioning. Fluorine is found in bones and teeth.

chapter 12

Nucleobase, Nucleoside, Nucleotide and Nucleic Acids

The nucleobases are purines and pyrimidines. There are two purines, namely, adenine and guanine and pyrimidines are thymine, cytosine and uracil. Adenine, guanine, cytosine and thymine are present in deoxyribonucleic acid (DNA) and ribonucleic acids (RNA) contain adenine, guanine, cytosine and uracil in place of thymine. Hence, one of the differences between DNA and RNA is in the presence of the nitrogen base, namely, the DNA has thymine while the RNA has uracil. All the other three bases are common to both DNA and RNA

The purines have a 9-membered ring with one benzene and another heterocyclic ring. The ring contains 4 nitrogen atoms. Here, the numbering of the atoms is anti-clockwise. The pyrimidines have a 6-membered ring and contains 2 nitrogen atoms.

STRUCTURE OF NUCLEOBASES

Adenine is 6-amino purine and guanine is 4-amino, 6-oxy purine. The cytosine is 2-oxy, 4-amino pyrimidine, uracil is 2, 4-dioxy pyrimidine and thymine is 2, 4-dioxy, 5-methyl pyrimidine.

A nucleoside is formed when a ribose or a deoxyribose is attached to the nucleobase by an N-glycosidic linkage between the nitrogen at the 9th position of purine and the hydroxyl group at the 1st carbon of pentose in a purine nucleoside. In a pyrimidine nucleoside, the N-glycosidic linkage is formed between the 1st nitrogen of the pyrimidine ring and the hydroxyl group at the 1st carbon of the pentose sugar. The nucleoside of adenine is adenosine, that of guanine is guanosine, that of cytosine is cytidine, that of thymine is thymidine and that of uracil is uridine. When the sugar attached to nucleobase is deoxyribose, then the nucleoside is deoxyadenosine, deoxyguanosine and deoxycytidine. Thymidine always contains deoxyribose and uridine invariably has ribose attached to it.

A nucleotide is formed when phosphoric acid is bound to a nucleoside, and forms adenosine monophosphate (AMP), guanosine monophosphate (GMP), cytidine monophosphate (CMP), thymidine monophosphate (TMP) and uridine monophosphate (UMP). Here again, the deoxy nucleotides shall be prefixed with the word "deoxy" and hence we have deoxy-AMP, deoxy-GMP and deoxy-CMP. The TMP always contains deoxyribose and UMP always shall have ribose. The phosphate is added to the hydroxyl group on 3rd or 5th carbon of the pentose.

The monophosphates of nucleosides can be sequentially phosphorylated to diphosphates and then to triphosphates resulting in ADP, ATP, GDP, GTP, CDP, CTP, TDP, TTP, UDP, and UTP. The phosphorylation is done by kinase enzymes.

The major function of nucleotides is to form the polynucleotides, DNA and RNA. The nucleotides also function as high energy compounds, for example, ATP and GTP. The hydrolysis of ATP to ADP releases 7.4 kilo calories of energy. This supplies the energy for several energy requiring reactions of the body. The GTP is required for protein synthesis along with ATP. The UTP as UDP-glucuronic acid is required for conjugation of free bilirubin and UDP-glucose is used for glycogenesis. It is also needed for galactose synthesis. CDP-choline is involved in the synthesis of phospholipids.

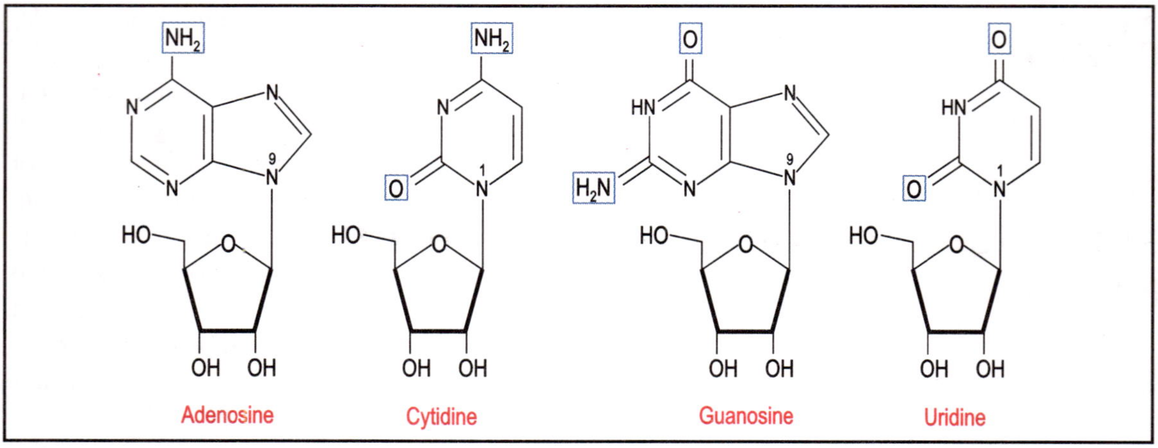

Fig. 12.1: Ribonucleosides

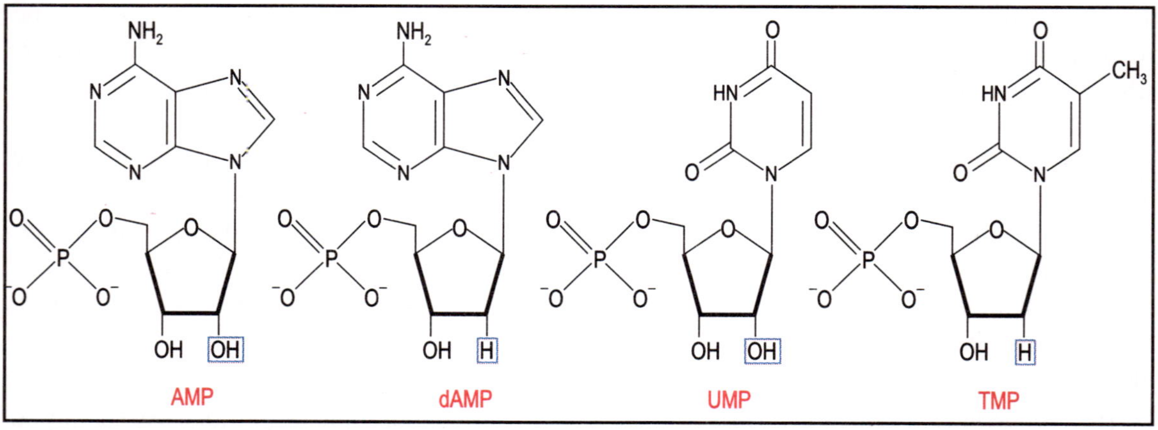

Fig. 12.2: Structure of AMP, dAMP, UMP, and TMP

The coenzymes contain nucleotides in their structure, for example, NAD, NADP, FAD, FMN, TPP, coenzyme A and vitamin B_{12}.

There are cyclic nucleotides like Cyclic AMP and Cyclic GMP. The Cyclic AMP is formed from ATP in presence of the enzyme adenyl cyclase, which is activated by glucagon and epinephrine. The Cyclic AMP acts as a second messenger through which several hormones act. The Cyclic GMP is also a mediator of hormone action.

The cyclic AMP is destroyed by phosphodiesterase which is inhibited by caffeine and theophylline. So excess intake of caffeine is not good. As it may cause lipolysis.

The nucleoside analogues are employed as anticancer agents. The examples include 5-fluorouracil, 5-iodouracil, azaguanine, cytarabine, etc. These, during replication, are incorporated into the growing DNA strand, but further growth of DNA is arrested and hence cell division is stopped. Other mechanisms for the generation of toxic effects may be due to the anti-cancer effect of these analogues on various metabolic processes.

NUCLEIC ACIDS

The nucleic acids are DNA and RNA. DNA is present in the nucleus and mitochondria. The RNAs are of different types like messenger RNA, ribosomal RNA, and transfer RNA. Other varieties of RNAs are hetero- nuclear RNA, small nuclear RNA, micro-RNAs and interfering RNAs.

Deoxyribonucleic Acid

The DNA is a polymer of deoxyribonucleotides and exists as a double, stranded molecule very much like a spiral stair case, with the two strands having opposite polarity. One strand runs from 5′ to 3′ while the other strand runs from 3′ to 5′, that is they are antiparellel in nature. The DNA is a double helical molecule and each turn of the helix contains 10 base pairs and has a pitch of 3.4nm. The nucleotides in the same strand are linked by 3′ to 5′ phosphodiester bonds which form the handrail of the stair case. The nucleotides of one strand is linked to the complementary nucleotides of the opposite strand by hydrogen bones. For example, adenine of one strand is hydrogen bonded to thymine of the opposite strand and guanine of one strand is linked to the cytosine of the opposite strand. The nucleotides form the footsteps of the staircase. Adenine and thymine are known as complimentary bases, so are guanine and cytosine. This is known as base pairing rule. In a DNA molecule the concentration of adenine is always equal to thymine. The same is also true for guanine and cytosine which are also equal in DNA. The DNA also can be envisioned as two parallel roads carrying traffic in opposite directions. The DNA has two grooves winding across the length, one is a minor groove and the other a major groove. Proteins like histones and nonhistones bind at these grooves forming a protein coat and regulating the expression of genes. One strand of the DNA is the template (sense) strand and the other strand is called as anti-template (anti-sense) strand. The mRNA formed is complimentary to the template strand and the same as anti-template strand with the sole change of uracil replacing (in RNA) thymine of DNA. The template strand is also designated as non-coding strand and the anti-template strand is called as the coding strand. In a DNA with several genes, it is not necessary that all the genes have the same strand as the template strand.

Diagram of structure of DNA

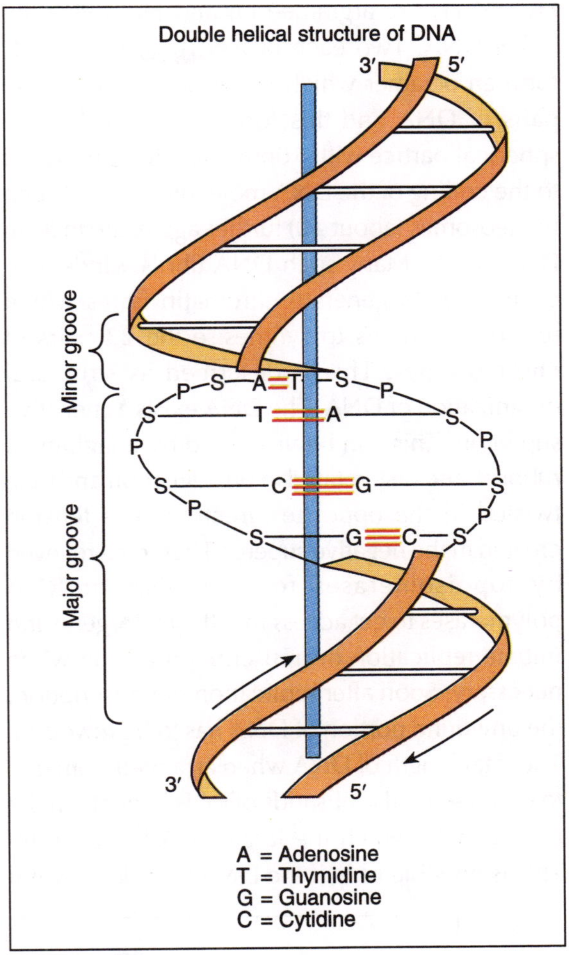

Fig. 12.3: Double helical structure of DNA

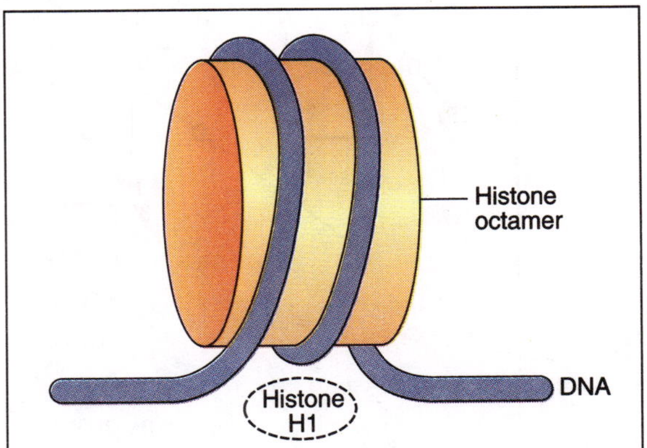

Fig. 12.4: Nucleosome structure

The histones are of 5 types. They are H_1, H_{2A}, H_{2B}, H_3 and H_4. The H_{2A} and H_{2B} are lysine-rich, the H_3 and H_4 are arginine-rich and H_1 covers the linker DNA. Two each of H_{2A}, H_{2B}, H_3 and H_4 form an octamer which covers almost 186 base pairs of DNA and this forms a nucleosome, a spherical partice with 10nm diameter. This leads to the coiling of the DNA molecule. Several such nucleosomes (about 10) further aggregate to form DNA fibrils. Many such DNA fibrils winds over each other to generate chromatin fibres which further supercoils to synthesise the 23 pairs of chromosomes. This is described as structural organization of DNA. The DNA exists as negative supercoil. This can be visualised by imagining a rubber band first, rotated in one direction and then twisted in the opposite direction. The tension created in the negative supercoil has to be relieved by topoisomerases for the DNA or RNA polymerases to get access into the DNA/gene and initiate replication or transcription as and when necessary. Soon after replication or transcription, the unwound portion of DNA has to be rewound. The total length of DNA when arranged from end to end is several thousands of folds larger than the nucleus into which the large DNA has to fit in. This is possible only if the DNA is folded several times, and in fact, the DNA is folded about 8000 times.

During replication, newly formed strand of DNA grows from 5′ to 3′ end, antiparallel to the parent strand.

The degradation of DNA is done by deoxyribonu-clease enzymes, which may be endonuclease, acting from the inside of the molecule, or exonucleases acting from the outside of the molecule. Another group of enzymes, the restriction endonucleases, of which over 800 types have been discovered, can recognise specific sequences on the DNA and cleave the molecule at these sites. The sequences recognized are known as palindrome sequences, which reads the same in both the strands, in the forward and backward directions. These enzymes can result in sticky ends or blunt ends.

Palindrome sequence

5′——AATT——3′
3′——TTAA——5′

Fig. 12.6: Plaindrome sequence in DNA (Cleavage site of restriction endonuclease)

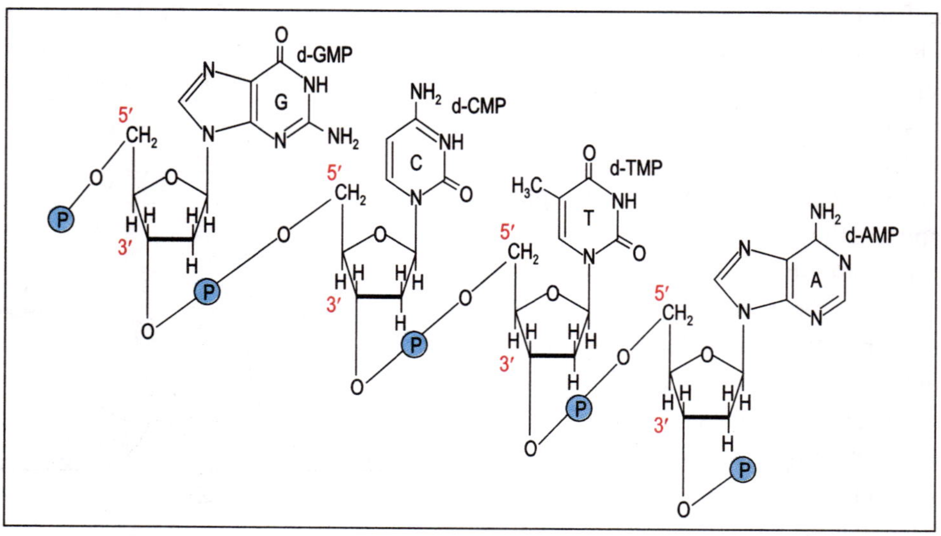

Fig. 12.5: Linkage of nucleotides in a single strand of deoxyribonucleic acid

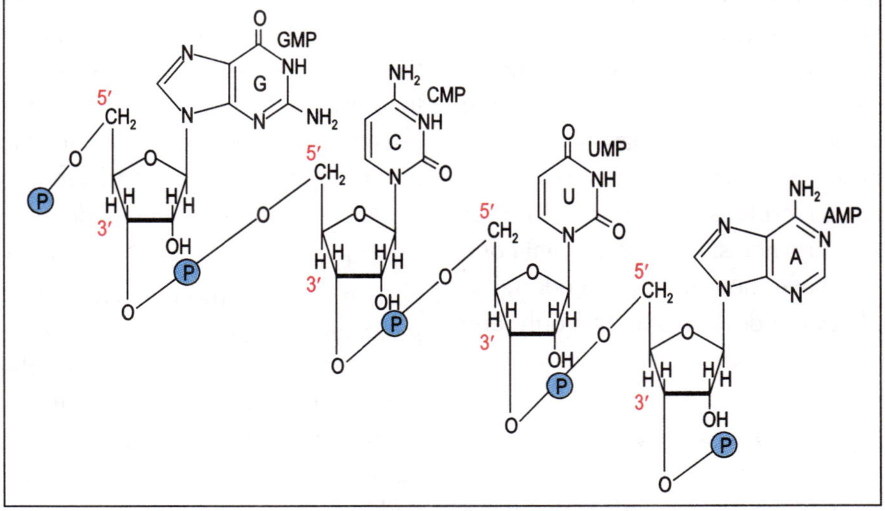

Fig. 12.7: Structure of ribonucleic acid

Ribonucleic Acids

The ribonucleic acids are polymers of ribonucleotides and are single-stranded structures. The pentose sugar present is ribose and the nucleobases present are adenine, guanine, cytosine and uracil with no thymine present. As the RNA is a single stranded molecule, it is not necessary that the amount of guanine is always equal to cytosine or that of adenine is always equal to that of uracil. Here again, the nucleotides are linked by 3′5′-phosphodiester linkages, as is true for DNA also. As the RNA is formed from DNA, it will have an opposite polarity to that of the template strand of DNA. The synthesis of RNA commences from 5′ to 3′ end in the process of transcription, reading the template strand of DNA from 3′ to 5′, by DNA dependent RNA polymerase. The nucleotide sequence of mRNA is complimentary to the template (non-coding) strand of DNA but is similar to the anti-template (coding) strand of DNA with the exception of uracil replacing thymine in the RNA.

The different types of RNAs include messenger-RNA (mRNA), ribosomal RNA (rRNA) and transfer RNA (tRNA). Additionally, there are other types of RNAs like hetero-nuclear RNA, small nuclear-RNA, micro-RNA and RNAi.

Messenger RNA

The messenger RNA carries the message of genes, in the form of triplets of nucleotides, from the template strand of DNA. The mRNA has many special features like a 'cap' of 7-methyl GTP at the 5′ end and a poly Adenosine 'tail' at the 3′ end. Between these two ends the triplet nucleotides, known as genetic code, are arranged. The genetic code is a triplet made up of three nucleotides. The deciphering of the genetic code took years of experiments in which the Indian born scientist Hargobind Khorana gave immense contribution by preparing polynucleotides of the same triplet and translating this message into amino acid sequences of a polypeptide. By this method it was found out which genetic code represents which amino acid. As there are only 4 nitrogen bases to code for 20 amino acids, and, if we presume that one base is coding for one amino acid, we can code for only 4 amino acids. If we presume, the code is made up of 2 bases, then a total of 16 amino acids can be coded. A total of

64 code words can be present if the code is a triplet. By experiments, it is irrevocably confirmed that the genetic code is a triplet and out of the 64 codons, 61 codes for one amino acid or other, the remaining 3 codons function as chain termination codons or nonsense codons. That means specific codons out of 61 codons, are responsible for coding for 20 amino acids. It is also true that one amino acid can have more than one codon, but one codon always codes for one and only one amino acid.

Properties of genetic code

The first property of the genetic code is that it is a triplet, consisting of three nucleotides.

The genetic code is unambiguous which means that, given a genetic code, it always codes for only one amino acid, for example the codon AGA codes for the amino acid, arginine and the codon UUU codes for the amino acid phenyl alanine.

Although, the genetic code is unambiguous, it has also got the property of degeneracy, meaning, that one amino acid can be indicated by more than one codon. This is due to the lack of stringency in the binding of the third nucleotide of the codon on the mRNA with the corresponding nucleotide of the anticodon on the tRNA. This means that the third nucleotide of the codon of mRNA is less strict and can be altered without any change in the amino acid incorporated into the protein during the process of translation. This is also referred to as wobble phenomena which means that this nucleotide to nucleotide attachment can wobble and is flexible. For example, the amino acid arginine can be called by the codons AGA and AGG. Therefore, certain amino acids have more than one code, sometimes up to 6 as in the case of serine. There are certain other amino acids, like methionine and tryptophan having only single code. This shows, in general, that the first two nucleotides in a triplet codon are more important than the third nucleotide in specifying the amino acid to be incorporated into the proteins. This further tells us that a change in the third nucleotide of a codon, in general, does not affect the accuracy of amino acid fabrication in the protein molecule and can account for the degeneracy of the genetic code.

The genetic code is non-overlapping and the protein synthesising system reads the messages in separate entities, without any overlapping between the codes. Further, when once the reading is started, it is read in a continuing sequence of nucleotide triplets without any punctuations till the system meets with a translation termination codon.

With a few exceptions, the genetic code is considered to be universal, meaning, the same genetic code codes for the same amino acid in all the living organisms, barring a few exceptions like certain codons in the mitochondrial mRNA being read differently from that those of cytoplasmic mRNA. In mammalian mitochondria, AUA and UGA are read as methionine and tryptophan codons respectively. These codons in cytoplasmic translation serve as codons for methionine (AUA) and chain termination (UGA). The codons AGA and AGG are read as stop codons in the mitochondria, rather than for arginine in cytoplasmic translation process. Therefore, the genetic code can be considered to be universal if the few exceptions, noted, are excluded.

Transfer RNA

The transfer RNAs have the shape of a clove leaf with different loops or arms with specific functions. The arms include anticodon arm, acceptor arm, thymidine-pseudouridine-cytidine arm for fixing the amino-acyl tRNA on the ribosomes and a dihydro-uracilo arm for the identification of the tRNA. The anti-codon arm dictates the amino acid to be bound to the acceptor arm. A specific transfer-RNA (tRNA) functions to bind to a specific

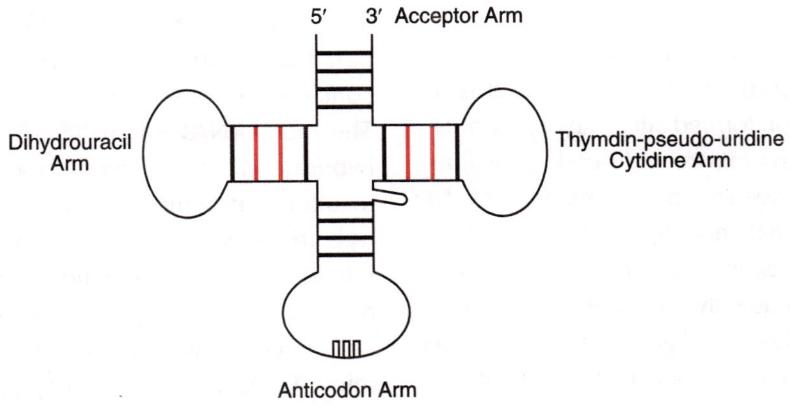

Fig. 12.8: Structure of t-RNA

amino acid and transports the amino acid to the site of protein synthesis, the ribosomes. There should be a minimum of 20 types of tRNAs, as there are 20 amino acids, one for each amino acid. The reaction between tRNA and its specific amino acid is highly specific and depends on the anti-codon of the tRNA. Although, 20 tRNAs are sufficient for cytoplasmic and mitochondiral translation, in fact, there are 32 tRNAs present in cytoplasm as against 20 present in mitochondria.

The binding of amino acid to its specific tRNA is known as charging of tRNA. This requires the specific amino acid, its tRNA, ATP as a source of energy and the specific enzyme, amino-acyl-tRNA synthetase. These enzymes are specific for specific amino acids and its corresponding tRNAs.

Amino acid + ATP + Enzyme ⟶ [Amino acid – AMP – Enzyme] + ppi

$$\text{[Amino acid – AMP – Enzyme] + tRNA} \xrightarrow{\text{Amino-acyl-tRNA synthetase}} \text{amino-acyl-tRNA + AMP + E}$$

Hence, the activation of amino acid prior to its role in the protein fabrication, consumes 2 ATP moles, as 1 ATP is hydrolysed to AMP and ppi. These enzymes are specific for an amino acid and its specific tRNA.

The t-RNAs are relatively small ribonucleic acids, about 75 nucleotides, also have a polarity and at the 3′ end have a CCA sequence and the hydroxyl group adenosyl moiety, and the carboxyl group the amino acid is linked through an ester bond. Further, the tRNA moles are the adaptor molecules, recognizing a specific amino acid on one hand and a genetic code on the other, thereby, warding off the remotest chance of an error during the process of translation.

Ribosomal RNA

The ribosomal RNAs (rRNA) are associated with the ribosomes, the ribonucleoprotein particles, consisting of a larger subunit (60s and a smaller subunit, 40s) in the eukaryotes making a whole ribosome of 80s. In the prokaryotes the subunits are smaller, that is, 50s and 30s and subunits making a whole ribosome of 60s. The larger subunit consists of 5s, 5.8s and 28s rRNA and about 50 polypeptide chains. The small subunit contains a 17s rRNA and 30 polypeptide chains. When there is no protein synthesis taking place, the two subunits remains separate, but while translation process is on, the two subunits combine together to form the whole ribosome.

Hetero-nuclear RNA

This is the precursor of mRNA. The Hn-RNA is much larger than the mature mRNA and is the primary transcript, formed after transcription. The Hn-RNA contains introns (non-coding regions), interspersed between exons (coding) regions. The post-transcriptional modifications remove these introns, and the exons are joined by the process of splicing. Subsequently, a poly 'A' tail is attached at the 3′ end and 7-methyl guanosine triphosphate is added at the 5′ end as a cap so as to make the mRNA.

The splicing process is done by small nuclear RNA molecules, called as *small ribonucleoprotein particles* (snurps). They are ribozymes (RNA containing enzymes such as U_1, U_2, U_5 and U_6. The micro- RNAs are small RNA moles which can hybridise with complementary sequences on DNA to affect the gene expression. They can also hybridise with mRNA molecules, thereby interfering with the function of the mRNA, and hence the translation process. They are also known as interfering RNA or RNAi. It is possible to use micro-RNAs having base complimentarily with viral mRNAs to stop the translation of viral mRNA, thereby resulting in the control of viral infection. This can open new research areas for antiviral therapy.

chapter 13

Metabolism of Nucleotides and Gouts

Biosynthesis of Purine nucleotides

Every cell of the body except the brain and erythrocytes has got the capacity to synthesise the purine nucleotides de novo. The compounds contributing for the synthesis of purines are: The glycine amino acid donates C_4, C_5, and N_7 of purine. The N_1 nitrogen is from asparagine, N_3 and N_9 are from glutamine amino acids. The C_6 comes from respiratory carbon dioxide, whereas C_2 is from formyl tetrahydrofolate and C_8 from methenyl tetrahydrofolate.

Diagram to show the source of atoms of purine nucleotide:

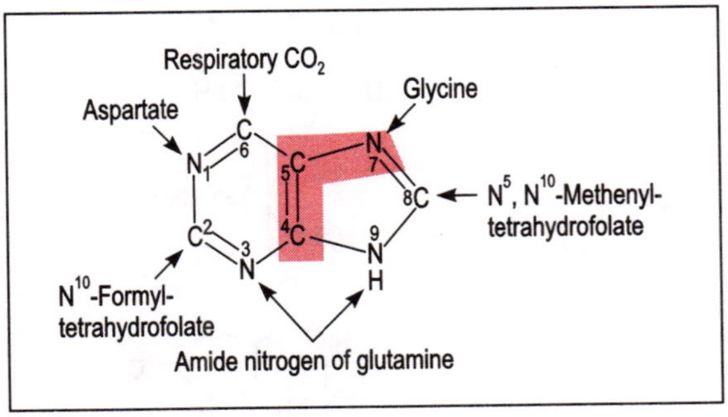

Fig. 13.1: Sources of atoms in purines

The first reaction serves to produce phosphoribosyl pyrophosphate (PRPP). The enzyme required is PRPP synthase. The substrates are ATP and ribose 5-phosphate.

$$\text{ATP} + \text{ribose-5 phosphate} \xrightarrow{\text{PRPP synthase}} \text{PRPP} + \text{AMP}$$

This is not the committed step in the biosynthesis of purine nucleotides as the PRPP is serving for the synthesis of several other compounds like pyrimidines, coenzymes like NAD, NADP, FAD, FMN, TPP, CoASH, etc.

The next reaction involves PRPP and glutamine to form 5-phospho ribosyl amine in presence of the enzyme amido transferase.

Both PRPP synthase and amido-transferase are subjected to feed back inhibition, the inosine monophosphate (IMP) acting as an inhibitor. At times, the genes of these enzymes are mutated and as a result of it, these enzymes may become feedback resistant. This leads to overproduction of PRPP, which in turn, leads to the formation of more purine nucleotides which are catabolised to uric acid causing hyperuricaemia and gout.

The 5-phosphoribosyl amine gets converted to IMP (inosine monophosphate) by a series of reactions. The IMP can then generate the other purine nucleotides, GMP and AMP.

The regulation of de novo purine synthesis is done by feedback inhibition of two enzymes, PRPP synthase and amidotransferase by AMP, GMP and IMP.

Purine salvage pathways

Certain tissues like brain do not possess the enzymes for the de novo synthesis of purine nucleotides and therefore, have to depend on readymade purines like hypoxanthine, guanine and adenine to be synthesised into IMP, GMP and AMP. The enzyme, hypoxanthine guanine phosphoribosyl transferase (HGPRTase) catalyses this conversion.

$$\text{Hypoxanthine} + \text{PRPP} \xrightarrow{\text{HGPRTase}} \text{IMP} + \text{PPi}$$

$$\text{Guanine} + \text{PRPP} \longrightarrow \text{GMP} + \text{PPi}$$

$$\text{Adenine} + \text{PRPP} \xrightarrow{\text{APRTase}} \text{AMP} + \text{PPi}$$

A deficiency of HGPRTase enzyme causes Leish Nyhan syndrome, characterised by neurological problems like self-mutilation and hyperuricaemia, gout and nephrolithiasis. The neurological problem is due to lack of purine nucleotides in the brain as a result of defunct purine salvage pathway.

Synthesis of deoxyribonucleoside diphosphates

Reduction of the 2′-hydorxyl of purine and pyrimidine ribonucleotides by the enzyme ribonucleotide reductase complex forms deoxyribonucleoside diphosphates (dNDPs) which can be further phosphorylated to dNTPs, used for the synthesis of DNA. The reductase enzyme is active in cells while involved in the synthesis of DNA. This reduction process requires thioredoxin, thioredoxin reductase and NADPH and NADPH-thioredoxin reductase.

Purine catabolism

The purine catabolism produces uric acid, a sparingly soluble waste product in comparison to soluble urea, the end product of protein catabolism.

In the first step of purine nucleoside catabolism, adenosine is deaminated to inosine by the action of the enzyme adenosine deaminase (ADA).

$$\text{Adenosine} \xrightarrow{\text{Adenosine deaminase}} \text{Inosine} + NH_3$$

In the second step, inosine gets transformed into hypoxanthine by the enzyme purine nucleoside phosphorylase in presence of inorganic phosphate. The by-product of the reaction is ribose-5 phosphate.

$$\text{Inosine} + \text{Pi} \xrightarrow{\text{Purine nucleoside phosphorylase}} \text{hypoxanthine} + \text{ribose-5 phosphate}$$

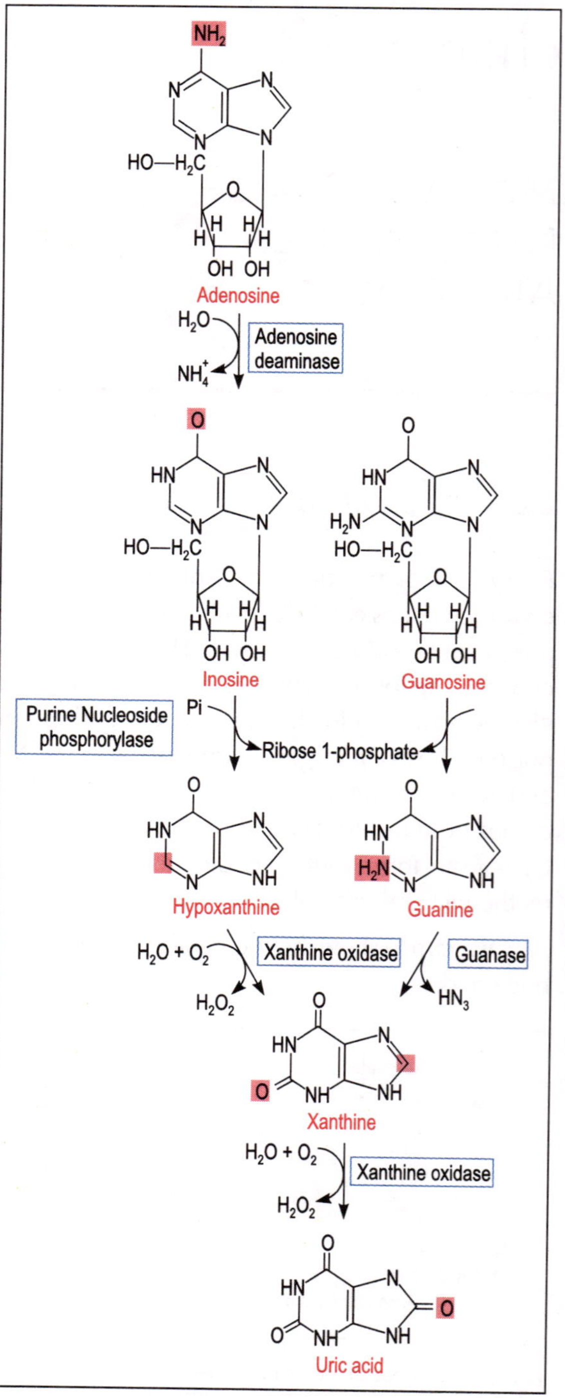

Fig. 13.2: Catabolism of purines

The third process involves the oxidation of hypoxanthine to xanthine by the enzyme xanthine oxidase which requires molybdenum as a cofactor and molecular oxygen. Hydrogen peroxide is also produced in this step.

In the fourth reaction xanthine is further oxidised to uric acid.

The second purine nucleoside, guanosine is converted to guanine by purine nucleoside phosphorylase enzyme. The guanine, with the help of guanase becomes hypoxanthine which subsequently forms uric acid via xanthine as described earlier.

Hyperuricaemia and gout

The normal serum concentration of uric acid is 3-7mg/dl in males and 2-5mg/dl in females. Hyperuricaemia can result from the overproduction of uric acid due to the abnormalities in the enzymes catalysing the synthesis of purine nucleotides like PRPP synthase and 5-phosphoribosyl amidotransferase both of which become feedback resistant due to mutations in the corresponding genes. Because of this, there will be a rise in the levels of PRPP which go for the synthesis of purine nucleotides, which accumulate, and hence, get catabolised to uric acid, resulting in hyperuricaemia.

The gout is a condition in which uric acid crystals are deposited in the synovial fluid of joints like the metatarsophalangeal joints causing inflammation and swelling accompanied by excruciating pain. Chronic condition, if persists, can lead to gouty arthritis and can cause destruction of the joints. Uric acid can accumulate in soft tissues, say in kidneys, causing nephrolithiasis. The solubility of uric acid is least in the cooler areas of the body.

Uric acid is only sparingly soluble in aqueous medium and exists in two forms: sodium urate, the salt form and uric acid, the acid form. The salt form is several times more soluble than the acid form. The pH of the medium determines the pK value of uric acid and which type of uric acid predominates. In alkaline condition, the more soluble sodium urate will be more, whereas in an acidic medium, the least soluble uric acid is more. Nephrolithiasis occurring after acidification of the urine will be of uric acid type, while that formed before acidification will be of sodium urate type.

Therefore, in alkaline state, the predominant salt form, being more soluble, can be excreted better by the kidneys. So, beneficial effects can be seen when we try and alkalinise the urine, say, by consuming citrate fruits.

The gout can be acute or chronic. Acute attacks of gout can be precipitated by alcohol consumption. The long standing gout causes destruction of the joints and is called chronic gout.

The gout is classified into primary and secondary gout. The idiopathic gout is one with no known etiology.

Primary gout

This condition arises due to abnormalities in the activity of the enzymes involved in the synthesis of purine nucleotides.

The PRPP synthase enzyme, though not a committed step in the purine nucleotide synthesis, as the PRPP is required for several other processes like formation of coenzymes, pyrimidines, etc., it is under the feedback control of IMP, AMP and GMP. It is found that certain mutations in the PRPP synthase gene can lead to the enzyme becoming less susceptible to feedback inhibition, resulting in the enzyme becoming 'overactive', so to say. As stated before, the synthesis of PRPP increases inspite of less need for purine nucleotides. These excess nucleotides, which are not used are degraded to uric acid. This process culminates in hyperuricaemia and gout.

The first committed enzyme in the purine synthesis is 5-phosphoribosyl amidotransferase, which is also under the feedback control of the end product, IMP. Here again, the corresponding gene may undergo mutation and the abnormal gene can code for an "overactive" amidotransferase and the fallout of this would be increased production of purine nucleotides and their subsequent catabolism to uric acid.

Thirdly, the deficiency of the purine salvage pathway enzyme, HGPRTase can indirectly lead to the accumulation of PRPP and produce more uric acid, resulting in gout. This is an x-linked recessive state and hence, encountered in males. The disease is known as Lesch-Nyhan syndrome. These patients exhibit neurological symptoms and self-mutilation, like biting nails, lips and inflicting pain to self. The neurological symptoms are due to the failure on the part of the brain to get necessary purine nucleotides through the savage pathway, as the brain lacks the pathway for the de novo synthesis of purines.

Another enzyme which, when absent, may cause hyperuricaemia is glucose-6 phosphatase, an enzyme required for both glycogenolysis and gluconeogenesis. This is type-1 glycogen storage disease (von Gierke's disease) in which glycogen, although present in plentiful in the liver, cannot contribute to blood glucose and hence, cannot be used for energy generation. The glucose-6 phosphate cannot form glucose due to the absence of the glucose-6 phosphatase enzyme. Therefore, the glucose-6 phosphate is diverted to HMP shunt pathway during which ribose-5 phosphate is formed. The ribose-5 phosphate can go for formation of PRPP, purines and uric acid.

The abnormality of glutathione reductase enzyme increases the demand for NADPH which further speeds up the HMP shunt pathway, producing more ribose-1-phosphate, PRPP, purine nucleotides and uric acid, in that sequence.

Secondary gout

The secondary gout includes the presence of malignancy, as the malignant cells divide faster accompanied by tissue destruction and catabolism of purines forming uric acid. Treating cancer with radiotherapy and/or chemotherapy also leads to tissue damage and hyperuricaemia.

Polycythaemia can also cause hyperuricaemia.

Consumption of alcohol in the night, by a patient of gout can aggravate his condition and he may wake up the next day morning with severe joint pain especially on the metatarsophalangeal region. On consumption of alcohol, it can be converted first to acetaldehyde and then to acetic acid, the enzymes catalysing these reactions being alcohol dehydrogenase and aldehyde dehydrogenase respectively. In each of the reactions 1 NAD molecule forms NADH. The raised NADH/NAD ratio favours the reduction of pyruvate to lactate. Hence, alcoholics suffer from lactacidosis. This causes an acid pH, resulting in the formation of the least soluble uric acid crystals, thereby worsening the condition. Further, the lactic acid competes with uric acid for excretion leading to the decreased excretion of uric acid. Hence, it is imperative that gouty patients should avoid taking alcohol.

Managing gout

The patient of gout should take more water so as to flush out uric acid. They should restrict, if not avoid non-vegetarian diet, which contains more nucleoproteins. They also should not consume alcohol for reasons explained previously. They should keep physically active.

The patients with gout can be given analgesic and anti-inflammatory medications. They can be given uricosuric drugs which increases excretion of uric acid.

Allopurinol acts as a competitive inhibitor of xanthine oxidase enzyme to prevent the formation of uric acid. Further, allopurinol is a suicidal inhibitor, meaning, the enzyme, xanthine oxidase acts on allopurinol to convert it to alloxanthine which becomes a more potent inhibitor of the enzyme. Therefore, allopurinol is often used in the treatment of gout.

ADA deficiency

Adenosine deaminase and purine nucleoside phosphorylase deficiency are associated with immunodeficiency states. The ADA deficiency causes the accumulation of dGTP and dATP which inhibit ribonucleotide reductase and decrease the levels of dCTP and TTP. Such a condition affects the replication process and hence, the division both B- and T-lymphocytes and causes severe combined immunodeficiency (SCID). The deficiency of the second enzyme purine nucleoside phosphorylase causes only T-cell defect, with normal B-cells, which is less severe.

Pyrimidine Metabolism

The pyrimidines, cytosine, uracil and thymine also can be synthesised in the body and takes place in the cytoplasm. The first step in the biosynthesis of pyrimidine is the reaction between CO_2, glutamine and ATP to form carbamoyl phosphate in presence of the enzyme, carbamoyl phosphate synthase-II (CPS-II), the CPS-I being used in the formation of urea, occurring in the mitochondria.

$$CO_2 + \text{Glutamine} + \text{ATP} \xrightarrow{\text{Carbamoyl phosphate synthase}} \text{Carbamoyl phosphate} + \text{ADP}$$

The second step involves the reaction between carbamoyl phosphate and aspartic acid in presence aspartate transcarbamoylase to generate carbamoyl aspartate.

$$\text{Carbamoyl phosphate} + \text{Aspartate} \xrightarrow{\text{Aspartate trans carbamoylase}} \text{Carbamoyl aspartate} + \text{Pi}$$

In the third step the carbamoyl aspartate is converted to dihydroorotic acid, catalysed by the enzyme dihydroorotase.

Dihydro-orotic acid is dehydrogenated to orotic acid by dihydro-orotate dehydrogenase in the subsequent step.

In the next reaction, orotic acid reacts with PRPP to form OMP in presence of the enzyme orotate phosphoribosyl transferase.

Next, OMP is decarboxylated by the enzyme orotylic acid decarboxylase, to form uridine monophosphate (UMP), which can be phosphorylated to UDP and then to UTP. The CTP can be formed from UTP in presence of glutamine which provides the amino group. Further UDP can be reduced to dUDP which is dephosphorylated to dUMP. The dUMP forms TMP by the addition of methyl group from methylene tetrahydrofolate, catalysed by the enzyme thymidylate synthase.

Methotrexate inhibits the dihydrofolate reductase enzyme and hence the active form of folate, the tetrahydrofolate, cannot be synthesised. Therefore, methotrexate can inhibit the formation of TMP from dUMP.

The regulation of pyrimidine biosynthesis is affected by controlling the first two enzymes, namely, carbamoyl phosphate synthase II and aspartate transcarbamoylase, inhibited by UDP and CTP, but activated by PRPP synthase.

The synthesis of pune and pyrimidine nucleotides is well coordinated on a mole-to-mole basis and hence, parallel to each other. The regulation across the two pathways is possible at the level of PRPP.

The catabolism of pyrimidines produces water soluble products like CO_2, ammonia, beta alanine and beta amino isobutyrate. As all these metabolites can be excreted from the body, these do not accumulate in the body.

Orotic aciduria can be present due to the deficiency of orotate phosphoribosyl transferase and orotidylate decarboxylase.

chapter 14

Tricarboxylic Acid Cycle and Electron Transport Chain

TRICARBOXYLIC ACID CYCLE

TCA cycle constitutes a series of reactions taking place in the mitochondria in which the reducing equivalents like NADH and $FADH_2$ are produced. It is the final common pathway for carbohydrate, protein and lipid metabolism. It is also an amphibolic pathway meaning it is both catabolic and anabolic. The anabolic part is for the synthesis of nonessential amino acids and glucose. The fatty acids are oxidised to acetyl CoA by beta oxidation in the mitochondria and enter the TCA cycle by combining with oxaloacetate to form citrate. In this reaction, catalyzed by citrate synthase, the 2-carbon compound acetyl CoA reacts with the 4-carbon compound oxaloacetate to form citrate. Similarly, glycolysis produces pyruvate which will be converted to acetyl CoA by pyruvate dehydrogenase complex and this can enter the TCA cycle. Several of the amino acids enter the TCA cycle at stages of pyruvate, alpha-ketoglutarate, succinyl CoA, fumarate and oxaloacetate. The hydrolysis of the thio-ester bond of acetyl CoA generates sufficient energy to drive the reaction forward. This reaction is irreversible and the opposite reaction is carried out by ATP-citrate lyase which splits citrate to acetyl CoA and oxaloacetate. The acetyl CoA is the precursor for fatty acid and cholesterol synthesis in the cytosol. It is also the precursor for ketone body synthesis in the mitochondria.

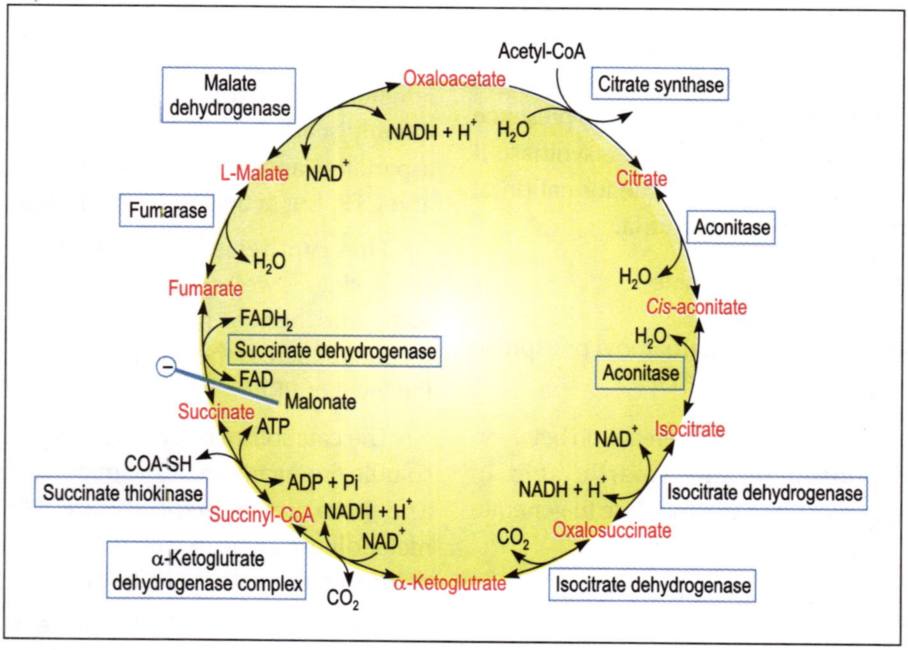

Fig. 14.1: TCA cycle

In the next step, citrate is isomerised to isocitrate through 2 reactions by the enzyme aconitase. The unstable intermediate is cis aconitate which is converted to isocitrate. The first reaction involves removal of water and the second addition of the same.

In the third step, the isocitrate is dehydrogenated to oxalosuccinate, the hydrogen removed from isocitrate being accepted by NAD which in turn becomes NADH. The oxalosuccinate, an unstable compound, gets decarboxylated to form alpha-keto gluterate. Both the above reactions are catalysed by the same enzyme, isocitrate dehydrogenase. The isocitric dehydrogenase has two isoforms, the mitochondrial one using NAD and the cytoplasmic one using NADP. The NADH can go to the electron transport chain with the liberation of 2.5 ATP moles.

The fourth step involves the oxidative decarboxylation alpha-ketoglutarate to succinyl CoA in presence of alpha ketoglutarate dehydrogenase complex, similar to PDH complex, and has three enzymes and 5 coenzymes. The NADH formed in this step, can be oxidised in the electron transport chain to produce 2.5 ATPs. A second molecule of CO_2 is released in this step. The enzymes required for this processes are (a) alpha ketoglutarate dehydrogenase, (b) dihydrolipoyl transsuccinylase and (c) dihydrolipoyl dehydrogenase. The coenzymes required are TPP, Coenzyme A, lipoamide, FAD and NAD.

The fifth step involves the only substrate level phosphorylation in TCA cycle and this happens when succinyl CoA is converted to succinate by succinate thiokinase. The energy released during the hydrolysis of the thio-ester bond of succinyl CoA is used to synthesise one GTP in presence of GDP and phosphate by the only substrate level in phosphorylation step of TCA cycle.

The sixth step is brought about by the enzyme succinate dehydrogenase which removes two hydrogen moles from succinate to convert it to fumarate. The hydrogen acceptor here is FAD which becomes $FADH_2$ that can generate 1.5 ATP moles in the ETC. The inhibition of succinate dehydrogenase by malonate serves as an example of competitive inhibition. The fumarate is an unsaturated acid with two carboxylic groups.

In the seventh reaction, L-malate is formed from fumarate by the enzyme fumarase and the reaction involves the addition of water.

The eight step leads to the regeneration oxaloacetate from malate and the enzyme catalyzing the reaction is malate dehydrogenase, with NAD being converted to NADH which can produce 2.5 ATPs. The oxaloacetate, thus regenerated, plays a catalytic role, as it is regenerated without being altered, and starts a new cycle by condensation with another molecule of acetyl CoA. The oxaloacetate is at the junction point of several pathways. In addition to initiating the citric acid cycle, the oxaloacetate can undergo transamination to give rise to the non-essential amino acid, aspartic acid. Further, it is also in the main pathway of gluconeogenesis.

In the whole of TCA cycle, three NADH and one $FADH_2$ moles are produced. The reactions that produce NADH in TCA cycle are isocitrate dehydrogenase, alpha ketoglutarate dehydrogenase and malate dehydrogenase. The $FADH_2$ is produced by succinate dehydrogenase. These reducing equivalents enter the electron transport chain and generates 9 ATPs. If we add up the one GTP (equivalent to ATP), produced by the substrate level phosphorylation, (at succinate thiokinase stage), the total number of ATPs produced in one turn of the reactions is 10.

The two carbons of acetyl CoA, obtained from glucose, fatty acids and amino acids, are both liberated as two CO_2 molecules, meaning complete oxidation of acetyl CoA occurring in one cycle, although the oxaloacetate is regenerated. The fact that both the CO_2 molecules generated in one turn of the cycle comes from acetyl CoA

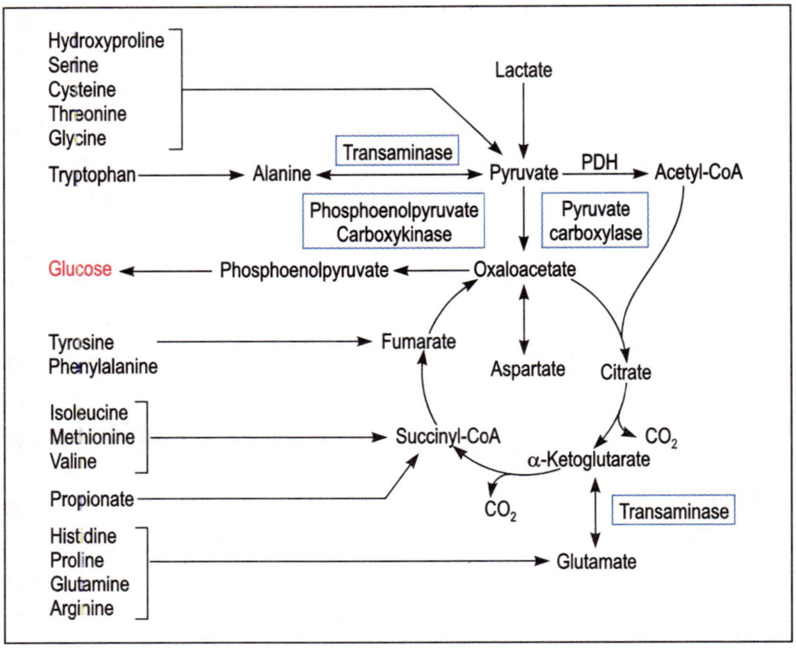

Fig. 14.2: Entry of amino acid carbon skeletons into TCA cycle

part of citrate and not from oxaloacetate can be proved by feeding radioactively labelled (C_{16}) fatty acids, in which case, the CO_2 liberated will be radioactively labelled.

The carbon skeletons of amino acids enter the TCA cycle at different points: Tryptophan and alanine can enter at pyruvate stage, aspartate at the level of oxaloacetate, phenyl alanine and tyrosine at fumarate level, valine, isoleucine and methionine enter TCA cycle at succinyl CoA level, and glutamate, histidine, arginine and proline enter at alpha keto glutarate stage.

The glucose is converted to pyruvate which becomes acetyl CoA that enters the Kreb's cycle. The even chain fatty acids are oxidised to acetyl CoA, whereas the odd chain ones produce acetyl CoA and the last 3-carbon compound remaining after beta oxidation is propionyl CoA which gets converted to methyl malonyl CoA. The latter forms succinyl CoA in presence of mutase, requiring adenosyl cobalamine.

As it is an amphibolic pathway, it is both anabolic and catabolic. All the amino acids that form the intermediates of TCA cycle can be converted to glucose. The even chain fatty acids cannot be converted to glucose, as they form acetyl CoA, which eventually is liberated as CO_2. Hence, it is not possible to get glucose from fat (even chain fatty acids) but fat can be synthesised from glucose. The fatty acids can be synthesised from acetyl CoA.

Further, the oxidation of fat requires small amounts of oxaloacetate to combine with acetyl CoA and this requires oxaloacetate mainly formed from pyruvate by pyruvate carboxylase enzyme. As the major source of pyruvate is glucose, there is an obligatory need for glucose for the oxidation of fat in the body.

Regulation

ATP is an allosteric inhibitor of citrate synthase, the regulatory enzyme of Kreb's cycle. Hence, the operation of TCA cycle is controlled by the energy (ATP) status of the cell. The TCA cycle is inhibited by NADH, but ADP activates it. The Kreb's cycle is coupled to ETC and oxidative phosphorylation.

The genetic abnormalities of TCA cycle are very rare and deficiency of any of the enzymes of the pathway is incompatible with life, as fat carbohydrate and protein metabolism require the TCA cycle.

ELECTRON TRANSPORT CHAIN AND OXIDATIVE PHOSPHORYLATION

The electron transport chain is located in the mitochondria. The mitochondria are the subcellular organelles which has two membranes, the outer membrane is permeable to a variety of molecules, whereas the inner membrane is only selectively permeable, the transport being done through specific transport molecules. The inner membrane contains cristae on which the electron transport chain (ETC) is situated and is associated with the oxidative phosphorylation.

The ETC consists of several electron carriers arranged from electronegative to electropositive components as electrons flow from electronegative to positive carriers. The ETC consists of 4 protein complexes. The first electron carrier in the chain is NADH which is produced by the reactions catalysed by the following enzymes: (1) glyceraldehyde-3 phosphate dehydrogenase (2) isocitrate dehydrogenase (3) malate dehydrogenase (4) beta hydroxy acyl CoA dehydrogenase and (5) glutamate dehydrogenase. In addition, pyruvate dehydrogenase and alpha keto glutarate dehydrogenase also produce NADH via lipoate and FAD.

The complex 1 consists of NADH, coenzyme Q, FMN and an iron sulphur protein. The electrons and hydrogen are transferred from NADH to FMN to produce NAD and $FMNH_2$. Next, the electrons from $FMNH_2$ is passed to iron sulphur protein and then to ubiquinone, which is also termed as coenzyme Q. Here the energy liberated is 12 kilo calories which drives protons to the outside of mitochondria. This energy is trapped as ATP.

The $FADH_2$ passes electrons to the ETC at the level of coenzyme Q, located in the complex 2. Here the calories liberated are not sufficient to remove the protons to the outside of mitochondria, meaning that $FADH_2$ bypasses the first complex, and hence, the first site of ATP generation. The enzymes producing $FADH_2$ are succinate dehydrogenase, acyl CoA dehydro-genase and glycerol-3-phosphate dehydrogenase. The $FADH_2$ passes the electrons to iron-sulphur protein which in turn transfers the electrons to ubiquinone. The ubiquinone accepts electrons from NADH (complex 1) and $FADH_2$ (complex 2). The ubiquinone has a long isoprenoid tail. No ATP is produced at this stage.

The complex 3 also contains iron sulphur proteins along with the haem proteins, cytochrome b and cytochrome c. Here the ubiquinone passes the electrons to iron sulphur proteins, which passes the electrons to cytochrome b, which further transfers the electrons to cytochrome c_1. The iron group in the cytochromes during oxido-reductions oscillates between ferrous and ferric states, unlike the haem group of haemoglobin in which the oxygen is only loosely bound to the haem of haemoglobin. The methaemoglobin having ferric atoms cannot perform its function. The free energy liberated can transport protons to the outside of mitochondria and 1 ATP can be produced at this point.

The complex 4 has cytochrome a and a_3. This complex is known as cytochrome oxidase which contains copper ions.

As per the new calculation, the oxidation of one mole of NADH in ETC produces 2.5 ATP moles and not 3 ATP moles, as thought early, and the oxidation of one mole of $FADH_2$ generates 1.5 ATPs and not 2 as thought before. Basing the calculation on the above facts, one mole of glucose

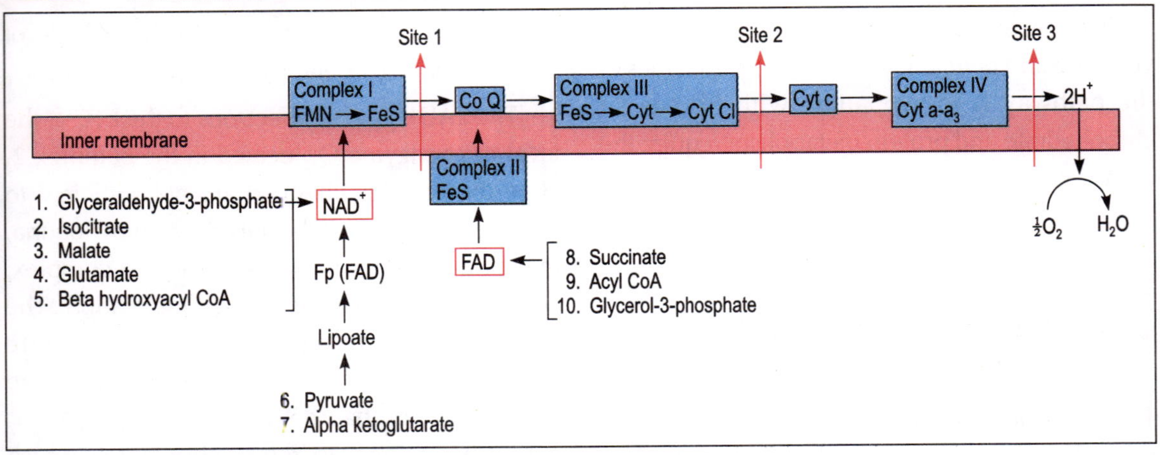

Fig. 14.3: Components and sequence of reaction of electron transport chain

generates 32 moles of ATPs, one mole of acetyl CoA forms 10 moles of ATPs and one mole of palmitate produces 106 moles of ATP.

Mechanism of oxidative phosphorylation

The coupling of oxidation and phosphorylation is termed as oxidative phosphorylation. The most accepted explanation of oxidative phosphorylation is put forward by Mitchel and this is known as chemiosmotic hypothesis. It is postulated that energy produced during oxido-reduction taking place in ETC leads to the liberation of energy between complex 1 and coenzyme Q (site 1), between cytochrome c_1 and c (site 2) and between cytochrome a-a_3 and molecular oxygen. The energy liberated is employed to drive the protons to outside of the mitochondria from inside, creating a proton gradient across the inner membrane. The protons accumulating outside the inner membrane causes the development of an electrochemical difference. With this the concentration of protons outside becomes more, and so, the protons enter the mitochondria through the F_0 serving as proton channels. The F_0 is connected to F_1 and the combination of both forms a transmembrane protein complex in the inner membrane of mitochondria, consisting of two functional units, the F_0 subunit, embedded in the membrane, serving as a proton channel and F_1 subunit which is a membrane protein projecting into the matrix, functioning as ATP synthase enzyme which forms ATP from ADP and phosphate.

With a decrease of ATP in the cells, the rate of TCA cycle and oxidative phosphorylation is increased. The reverse is true when the ATP levels increase.

The inhibitors of ETC and oxidative phosphorylation include cyanide (cytochrome a-a_3 inhibitor), CO, barbiturates, fish poison rotenone, insecticide, atractyloside, oligomycin valinomycin etc. he uncouples of oxidation and phosphorylation are dinitro phenol and dinitro cresol. The thyroxine in high doses and thermogenin in adipose tissue are considered as endogenous uncouplers of oxidation and phosphorylation. The uncoupling process assumes significance in newborn infants and in animals coming out of hibernation, in both of which, heat generation by thermogenesis plays a major role in adjusting to a cold environment. The newborn has to adjust to a cold extrauterine atmosphere in comparison to a warm intrauterine one. Here, the need of the hour is the generation of heat by uncoupling oxidation and phosphorylation.

The mitochondrial DNA is transmitted cytoplasmically, and hence, from the mother. It has several copies of the mitochondrial genes. The abnormalities in these genes can cause diseases of ETC and oxidative phosphorylation known as oxphos diseases, the examples of which are MELAS, myopathies, etc. The cytochrome c is implicated in the process of programmed cell death or apoptosis. The full form of MELAS is mitochondrial encephalopathy lactacidosis stroke like episodes.

chapter 15

METABOLISM— METABOLISM OF CARBOHYDRATES

METABOLISM

Metabolism consists of the several biochemical reactions going on in an organism including human beings. They include both the synthetic processes (anabolism) and the degradation reactions (catabolism). The energy liberated in the catabolic reactions is utilized for anabolic processes. The energy released during catabolism is trapped as high energy phosphate bonds of nucleoside triphosphates like adenosine triphosphate (ATP) or guanosine triphosphate (GTP). When we look at metabolism, it incorporates the reactions undergone by the major constituents of the body, like carbohydrates, proteins, lipids, nucleic acids, minerals, vitamins and water. The metabolism has to be studied in the context of health and disease. Hence, it is imperative that the study of metabolism during normal healthy (physiological) state should be preceding the alterations in the metabolism in different pathological scenarios. Biochemically speaking, health is that state in which all the biomolecules of the body are present in their correct structural conformation, thereby optimizing their functions with all the metabolic processes proceeding in a dynamic state, finely regulated and capable of responding to a variety of stimuli and stressful conditions, for example, an infectious state or a hormone imbalance. This is possible only with specific genes (DNA), getting transcribed into specific messenger–ribonucleic acids, to be further translated into proteins (products of gene expression) at the appropriate time, desired concentration and required rate. This would further entail the action of many an enzyme, coenzymes (vitamins), cofactors (minerals), secondary messengers and endocrines in a coordinated way.

We further can view metabolism through an organ based approach. For example, we may study the metabolism of liver, heart, skeletal, muscle, bones, brain, erythrocytes, etc. The approach then becomes system based.

We should never forget that although we may be tempted to group the metabolism into the sections, as mentioned, there is a lot of integration and overlapping between them. Hence, at a different level, we should not fail to have an integrated view of not only of the metabolism of various components, but of the whole organism, which is more true for the highest evolutionary organism, the humans. As a result, we should be in a position to understand what is health and how do we fall sick. This would open the doors for us to select specific laboratory investigations for the diagnosis of a variety of diseases. Such studies could further pave the way for the management of patients, including the formulation of newer drugs and testing their efficacy. With these thoughts, we would start to learn various aspects of metabolism in health and see how and why it alters in specific disease processes like diabetes, jaundice, gout, atherosclerosis, cardiovascular diseases and cancer to name a few.

METABOLISM OF CARBOHYDRATES

The major carbohydrates of the body include glucose, galactose, fructose, mannose, ribose, deoxyribose, maltose, lactose, glycogen and heteroglycans like chondroitin sulfate, hyaluronic acid, dermatan sulfate, keratin sulfate, heparin and ABO blood group substances and proteo-glycans. Out of these, we would talk more of glycogen and glucose in this session.

The carbohydrate metabolism is described to consist of digestion and absorption of carbohydrates, glycolysis, glycogenesis, glycogenolysis, gluconeogenesis, pentose phosphate pathway, metabolism of lactose, galactose, fructose, glucuronic acid pathway, blood glucose regulation, diabetes mellitus with metabolic alterations, acute and long term complications, glucose tolerance test and other tests, commonly employed to diagnose and manage diabetes mellitus.

Glycolysis

It is a process by which glucose is oxidized to pyruvic acid in the aerobic condition or, to lactic acid in the absence of oxygen. The glycolysis takes place in the cytoplasm as the necessary enzymes are all present in the cytoplasm. One glucose molecule, when oxidized, generates 2NADH moles which when oxidized in electron transport chain of mitochondria produces 5ATP molecules in the presence of oxygen. In the absence of oxygen the NADH generated cannot be oxidized in the electron transport chain, but is used to reduce pyruvate to lactate. There is substrate level phosphorylation taking place in both the presence and absence of oxygen. As the 6-carbon glucose is cleaved into two 3-carbon compounds (1, 3-bisphosphoglycerate) there are actually 2 substrate level phosphorylations which in turn generates 2 ATPs. In each of the first and third reactions of glycolysis, 1 ATP is utilized. Hence, the total number of ATPs produced in glycolysis is 9 in aerobic state and only 4 in anaerobic state but the net production is 7 and 2 in aerobic and anaerobic states respectively as 2 ATPs are used in the first and third steps of glycolysis. Insulin stimulates the process of glycolysis, whereas glucagon inhibits it.

Steps of glycolysis

In the first reaction, one high energy phosphate from ATP is transferred to the sixth carbon of glucose to form glucose-6-phosphate.

$$\text{Glucose} + \text{ATP} \xrightarrow{\text{Hexokinase/Glucokinase}} \text{Glucose-6-phosphate} + \text{ADP}$$

The reaction requires magnesium ions. This step is unidirectional and is catalysed by hexokinase and glucokinase. The hexokinase has a low K_m for glucose and its role is when the blood glucose level is low as is the case during fasting. Moreover, hexokinase can act on other hexoses like fructose and galactose also. In contrast the glucokinase has a high K_m and works when blood glucose levels are increased, as in a well-fed state. The hormone insulin activates glucokinase and as a resul elevates glycolysis. Being unidirectional in action, it is one of the key or rate-limiting enzymes of the whole pathway. It serves as a bottleneck, the passage through which can be regulated.

Nowadays the glucokinase reaction is increasingly being employed for the assay of plasma glucose in clinical laboratory as the enzyme specifically acts on glucose.

The second reaction involves the isomerisation of glucose-6-phosphate to fructose-6-phosphate. The enzyme for this process is phosphohexose isomerase which happens to catalyse the reaction in both ways. Stated otherwise, it is a reversible or a bidirectional enzyme.

In the third step, there is one more phosphorylation in presence of a second ATP, as fructose-6-phosphate becomes fructose 1, 6-biphosphate, catalysed by phosphofructokinase-1. ADP is formed from ATP after transferring phosphate to glucose. The reaction requires magnesium ions. Till now the pathway has consumed energy equivalent to 2 ATP moles.

Phosphofructokinase-1 (PFK-1)
Fructose 6-phosphate + ATP ⟶ Fructose 1,6-bisphosphate + ADP

The PFK-1 is also a unidirectional enzyme and, hence, is under the control of hormones, insulin stimulating it and glucagon inhibiting it. Therefore, this is the second regulatory step of glycolysis, the first being the one brought about by glucokinase. The PFK-1 is also an allosteric enzyme, the positive modulators are ADP and AMP and the negative modulator is ATP. The allosteric enzymes are those possessing another site to which activators (positive modulators) or inhibitors (negative modulators) bind to activate or inhibit it respectively.

The PFK-2 is another enzyme which converts fructose 6-phosphate to fructose 2, 6-bisphosphate, which, in turn, serves to increase glycolysis but inhibits gluconeogenesis. The velocity to substrate concentration graph of allosteric enzymes is sigmoid in shape. To get the K_m for such enzymes one has to employ the double reciprocal plot (1/v to 1/s).

In the fourth step, the fructose 1, 6-bisphosphate is cleaved in the middle by aldolase, an example of a lyase enzyme, to form 2 three-carbon units which are termed as glyceraldehydes 3-phosphate and dihydroxy acetone phosphate. Aldolase is a bidirectional or reversible enzyme like phosphohexoseisomerase.

Aldolase
Fructose 1,6-bisphosphate ⟷ Glyceraldehyde 3-phosphate + Dihydroxy acetone

Phosphate

In the next process the dihydroxy acetone phosphate is converted to glyceraldehyde-3-phosphate by phosphotriose isomerase which is a reversible enzyme.

Phosphotriose isomerase
Dihydroxy acetone phosphate ⟷ Glyceraldehyde-3-phosphate

During the next reaction the glyceraldehyde-3-phosphate is oxidized to 1,3-bisphosphoglycerate by the enzyme glyceraldehyde-3-phosphate dehydrogenase, in presence of NAD and inorganic phosphate. This reaction is also reversible. The NAD is reduced to NADH + H^+. The NADH can produce 2.5 ATP moles in the electron transport chain. As glucose gives rise to 2 glyceraldehyde-phosphates, 5ATP moles are generated in aerobic condition. In contrast the 2 NADH moles are used for reducing the two moles of pyruvate to lactate in the absence of oxygen.

Glyceraldehyde 3-phosphate dehydrogenase
Glyceraldehyde 3-phosphate + Pi + NAD ⟷ 1,3-bisphosphoglycerate + NADH + H^+

In the seventh step the 1, 3-bisphosphoglycerate is converted to 3-phosphoglycerate. First substrate level phosphorylation of glycolysis takes place in this step and the high energy phosphate at the first position of 1, 3-bisphosphoglycerate is transferred to ADP to form ATP. Phosphoglycerate kinase is the enzyme catalyzing this reaction. Further, this is a bidirectional reaction.

3-Phosphoglycerate kinase
1,3-bisphosphoglycerate + ADP ⟷ 3-Phosphoglycerate + ATP

As there are two 1, 3-bisphosphoglycerate moles present, two ATPs are formed at this level. Hence, the net production till now is 3 ATPs, as one ATP each was utilized in the first and third step.

In the subsequent step, 3-phosphoglycerate is isomerised to 2-phosphoglycerate by phosphoglycero mutase. Here the phosphate from the third carbon of glycerate is shifted to the second carbon.

Phospho glycerate mutase
3-Phosphoglycerate ⟷ 2-Phosphoglycerate

The eighth step involves the conversion of 2-phosphoglycerate to phosphoenol pyruvate by the enzyme enolase and this leads to the release of one water molecule with the formation of a high energy phosphate bond in phospho-enol-pyruvate. This is a reversible reaction. Fluoride inhibits enolase by complexing with magnesium ions which is required by the enzyme for its activity. When blood is drawn for glucose assay, fluoride containing tubes are employed because fluoride is an efficient inhibitor of glycolysis by removing magnesium. Failure to collect blood in fluoride containing bulbs can result in continuation of glycolysis in erythrocytes leading to false low plasma glucose levels, missing the diagnosis of diabetes mellitus.

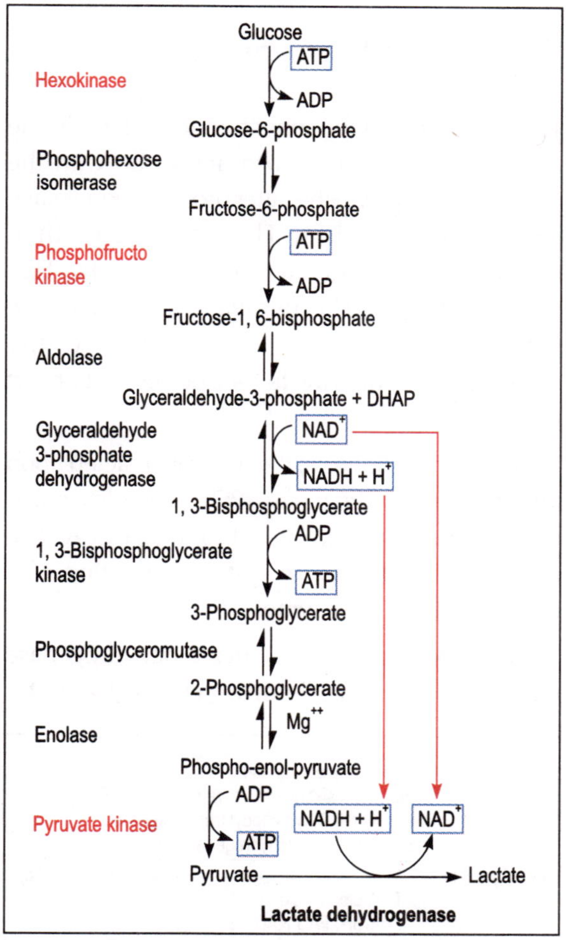

Fig. 15.1: Summary of glycolysis (Embden-Meyerhof-Parnas pathway

$$\text{2-Phospho glycerate} \underset{\text{enolase}}{\overset{Mg^{2+}}{\longleftrightarrow}} \text{Phospho enol pyruvate} + H_2O$$

In the succeeding process, phospho-enol-pyruvate is converted to pyruvate and an ADP is phosphorylated to ATP, the second substrate level phosphorylation in glycolysis. Here again 2 ATPs are formed as one glucose molecule generates 2 phospho enol pyruvate moles. Hence, in aerobic glycolysis the net number of ATP produced is 7 whereas the corresponding number in aerobic state is only 2, because, the 2 NADH produced are used for the reduction of pyruvate to lactate in the absence of oxygen. The enzyme involved is lactate dehydrogenase. This is specially important during exercise when the oxygen supply is limited and so the removal of NADH, essential for the continuance of glycolysis, through respiratory chain (ETC) is not possible. Therefore, the alternative to regenerate NAD, required for glycolysis, is to convert pyruvate to lactate. As the erythrocytes are devoid of mitochondria, the lactate production from pyruvate, coupled to the synthesis of NAD from NADH, attains significance as is also the case during exercise.

Glycolysis in erythrocytes

There are certain special features of glycolysis in erythrocytes due to the absence of mitochondria in these cells. The end product of glycolysis in erythrocytes is always lactate and the energy production is only 2 ATPs. The red blood cells need very little energy as there are no synthetic processes in it. They lack nucleus and other cellular organelles also. To maintain a low energy status, there is indeed an energy wasting step which is named as *Rappoport Luberring pathway*. This consists of conversion of 1, 3-bisphosphoglycerate to 2, 3-bisphosphoglycerate by bis phosphoglycerate mutase. The 2, 3-bisphosphoglycerate then becomes 3-phosphoglycerate by removing the phosphate at carbon number 2,

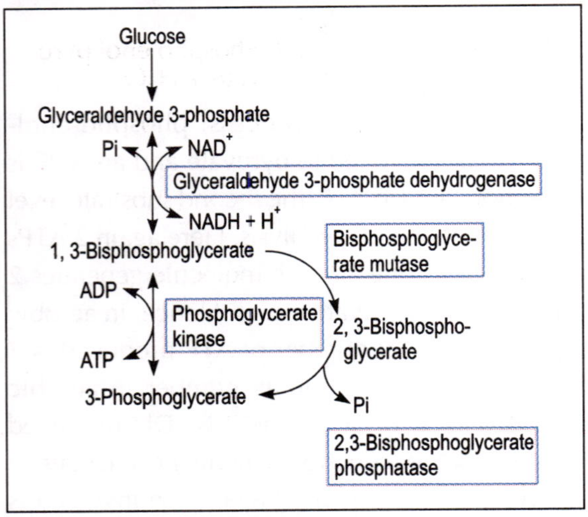

Fig. 15.2: Special features of glycolysis in RBC

catalysed by phosphatase enzyme. Here, as the phosphate bond at the second carbon is of low energy, there is no substrate level phosphorylation occurring. Hence, there is wastage of energy to the tune of 2 ATPs. Only 2 ATPs are formed in the erythrocytes which decreases the energy status of these cells, definitely helping in the continuance of glycolysis, essential for its survival.

Another significance of 2, 3-bisphopho glycerate (2, 3-BPG), produced only in RBCs is in the dissociation of oxygen from oxygenated hemoglobin in the tissues. If a blood with less 2, 3-BPG is transfused into patients, they can have hypoxia with life threatening consequences as the 2, 3-BPG-depleted blood cannot release oxygen. Hence the stored blood is mixed with adenosine, the ribose of which can maintain the 2, 3-BPG levels in the blood.

Regulation of glycolysis

The hormone insulin, secreted by the beta cells of islets of Langerhans of pancreas, in a well-fed condition, transports glucose from blood to tissues, stimulates glycolysis in addition to glycogenesis but inhibits gluconeogenesis and glycogenolysis, thereby preserving its role as a hypoglycaemic hormone. Opposite to insulin is glucagon, secreted by the alpha cells, which inhibits glycolysis and glycogenesis but stimulates gluconeogenesis and glycogenolysis, leading to a hyperglycaemic effect.

Insulin stimulates the key enzymes of glycolysis namely glucokinase, phosphofructokinase and pyruvate kinase. Glucagon has inhibitory effect on these enzymes.

Further, phosphofructokinase is subjected to allosteric stimulation by ADP and AMP but is allosterically inhibited by ATP. Fructose 2, 6-bisphosphate formed in a well-fed state simulates phosphofructokinase-1 resulting in increased rate of glycolysis.

Conversion of pyruvate to acetyl CoA

The pyruvate is now converted to lactate (in the absence of oxygen) and to acetyl CoA (in the presence of oxygen), the latter process taking place in the mitochondria. The acetyl CoA then complexes with oxaloacetate to enter the TCA cycle as citrate. The pyruvate, formed in the cytoplasm has to be transported to the mitochondria by pyruvate transporters to convert it to acetyl CoA.

In the mitochondria, the pyruvate is acted upon by pyruvate dehydrogenase (PDH) complex which oxidatively decarboxylates it to acetyl CoA. The PDH is a multi-enzyme system consisting of 3 enzymes and 5 coenzymes.

The first enzyme is pyruvate dehydrogenase which oxidatively decarboxylates pyruvate.

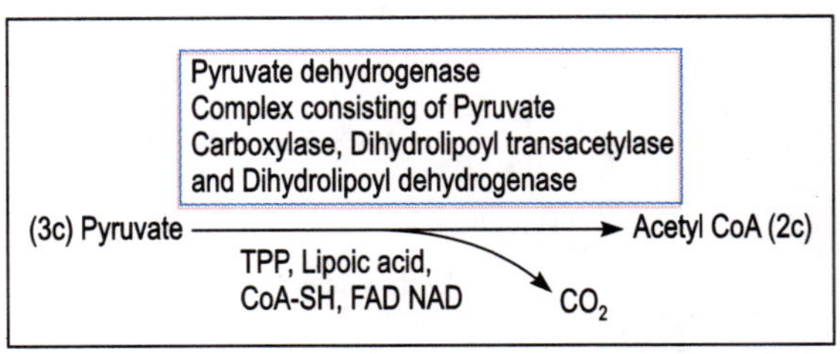

Fig. 15.3:

Thiamine pyrophosphate (TPP) is involved in this reaction and results in enzyme bound hydroxyethyl thiamine pyrophosphate with the concomitant liberation of carbon dioxide.

In the second step, the enzyme, dihydrolipoyl transacetylase, oxidises the hydroxyethyl moiety with subsequent attachment of this to the second coenzyme lipoamide to result in the formation of acetyl lipoamide. The acetyl group from acetyl lipoamide is given to coenzyme A which becomes acetyl CoA and the reduced form of lipoamide is also formed.

The third enzyme required is dihydrolipoyl dehydrogenase which helps in the conversion of reduced lipoamide to oxidised lipoamide, which involves the reduction of FAD to $FADH_2$. The oxidation of $FADH_2$ to FAD is coupled to the reduction of NAD to NADH + H^+. These reactions are similar to those involved in the oxidative decarboxylation of alpha ketoglutarate dehydrogenase to succinyl CoA in TCA cycle. The coenzymes involved in the two multi-enzyme complexes are TPP, lipoamide, COAsh, FAD and NAD.

The PDH complex is regulated by reversible phosphorylation, meaning PDH kinase phosphorylates PDH to make it inactive, whereas the PDH phosphatase dephosphorylates PDH to make it active. Further, acetyl CoA, NADH and ATP inhibit PDH. Higher levels of AMP will activate PDH.

The PDH is an irreversible enzyme, meaning the acetyl CoA cannot be converted back to pyruvate. Due to the lack of an enzyme capable of reversing this step, gluconeogenesis is not possible from even chain fatty acids, although the last three carbon-compound, propionyl CoA, resulting by the beta oxidation of odd chain fatty acids, can be processed to generate glucose.

The NADH formed during PDH action, generates 2.5 ATP moles while it passes through the electron transport chain. As there will be 2 pyruvate formed from 1 glucose molecule, the ATP formed will be 5.

The deficiency of thiamine in the diet can cause cessation of PDH action and can expectedly culminate in pyruvate excess and lactacidosis. Abnormalities in glycolysis can be detected resulting from the lack of enzyme, hexokinase, that may cause lowered levels of 2, 3 BPG, resulting in decreased oxygen dissociation. Contrary to this is the elevated concentrations of 2, 3, BPG prevalent in the deficiency of pyruvate kinase. The oxygen affinity for haemoglobin is less in the case of pyruvate kinase lack, whereas it is more when hexokinase is deficient. The pyruvate or for that matter, acetyl CoA are at the junction of various metabolic pathways. The pyruvate may be converted to oxaloacetate and then to glucose or get oxidatively decarboxylated to acetyl CoA, the pathway preferred being dependent on the nutritional and hormonal state of the individual. Pyruvate can be transaminated to alanin also.

GLUCONEOGENESIS

Gluconeogenesis is a process by which glucose is formed from noncarbohydrate sources like glucogenic amino acids, lactate, glycerol and the 3-carbon propionyl CoA, derived from the beta oxidation of odd chain fatty acids. The even chain fatty acids cannot be converted to glucose as the reaction catalysed by pyruvate dehydrogenase complex is irreversible. Further, both the carbons of the acetyl CoA formed by the beta oxidation of both even and odd chain fatty acids, after forming citrate by condensing with oxaloacetate, are liberated as 2 carbon dioxide moles in the TCA cycle at the level of isocitrate dehydrogenase and alpha-keto glutarate dehydrogenase.

Gluconeogenesis from lactate

Lactate, produced in the muscles during exercise, is taken to the liver for its conversion to glucose by Cori's cycle. For this, lactate is dehydrogenated by lactate dehydrogenase enzyme with the hydrogen removed from lactate being accepted by NAD which forms NADH + H^+. Meanwhile pyruvate gets reduced to lactate.

The various glucogenic amino acids undergo transdeamination to generate the corresponding keto acids which could enter the TCA cycle. Examples include alpha ketoglutarate, oxalo-acetate and pyruvate formed from glutamic acid, aspartic acid and alanine respectively. Asparagine can be deaminated to aspartate. Similarly glutamine, arginine, histidine and proline can give rise to alpha ketoglutarate. Isoleucine, methionine and valine enter the TCA cycle at succinyl CoA stage, so does the 3-carbon unit, propionate, remaining after the beta oxidation of odd chain fatty acids. The propionyl CoA has to become methyl malonyl CoA which has to be converted to succinyl CoA, the latter step requiring vitamin B_{12}. The deficiency of B_{12} causes methyl malonic aciduria. The amino acids, phenyl alanine, tyrosine, aspartate and asparagine can connect with TCA cycle at the stage of fumarate. Further, glycine, alanine, serine and cysteine can be trans-formed into pyruvate as is also the case with lactate.

There are 4 specific enzymes necessary for the conversion of pyruvate to glucose in addition to the reversible enzymes of glycolysis. The specific enzymes required for gluconeogenesis are pyruvate carboxylase, phospho-enol-pyruvate carboxy kinase, fructose 1, 6-bisphosphatase and glucose-6-phosphatase. The glucose-6-phosphatase is also involved in glycogenolysis. The glucogenic amino acids, after forming one of the intermediates of the TCA cycle, can get converted to oxaloacetate which can subsequently be converted to phospho-enol-pyruvate.

$$\text{Pyruvate} + CO_2 + \text{ATP} \xrightarrow{\text{Biotin / Pyruvate carboxylase}} \text{Oxaloacetate} + \text{ADP} + \text{Pi}$$

The pyruvate carboxylase, in presence of biotin, CO_2 and ATP converts the 3-carbon compound pyruvate to the 4-carbon compound oxaloacetate. It is an allosteric enzyme which is stimulated by acetyl CoA. The biotin combines with CO_2 to form carboxybiotin which then

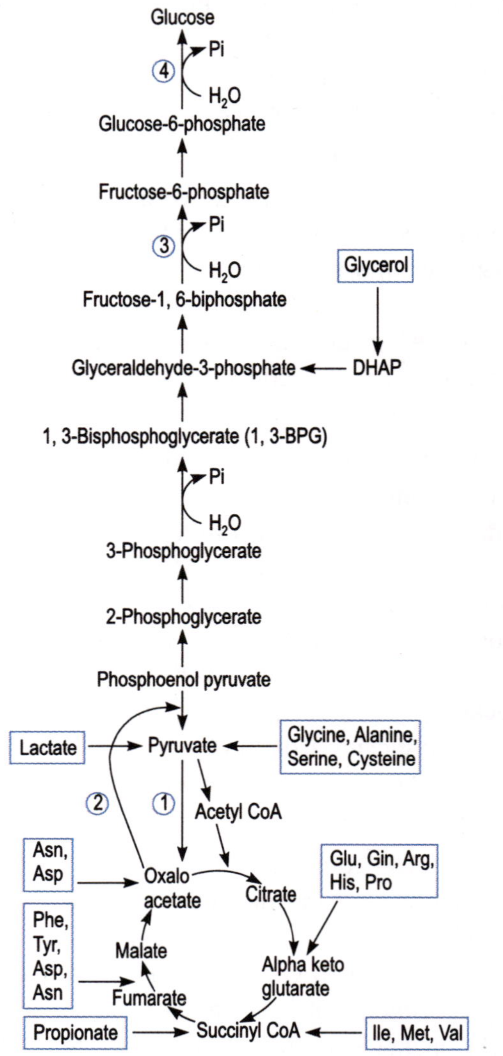

1. Pyruvate carboxylase; 2. Phosphoenol carboxykinase; 3. Fructose-1, 6-bisphosphatase; 4. Glucose-6-phosphatase

Fig. 15.4: Gluconeognic pathway

transfers the CO_2 to pyruvate to synthesise oxaloacetate. This enzyme is activated by glucagon and glucocorticoids but inhibited by insulin.

In the next step the oxaloacetate is transformed into phospho-enol-pyruvate by the enzyme phospho-enol-pyruvate carboxykinase (PEPCK). This reaction requires energy in the form of GTP and one mole of CO_2 is released. The enzyme PEPCK is also activated by glucocorticoids and

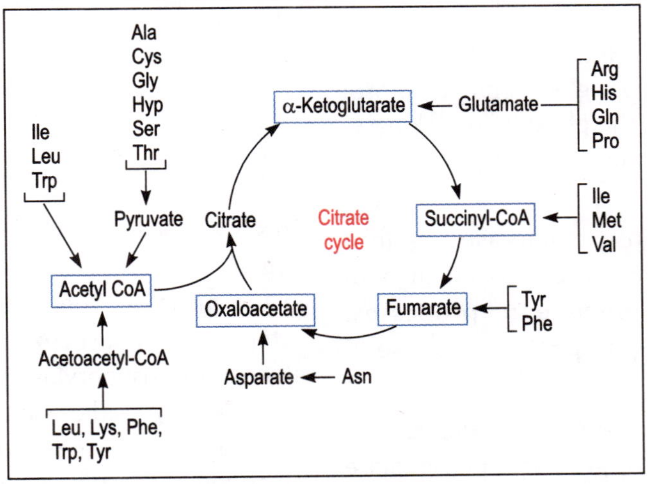

Fig. 15.5: Amino acid entry points to TCA cycle

glucagon and inhibited by insulin in a similar way as pyruvate carboxylase.

$$\text{Oxaloacetate + GTP} \xrightarrow{\text{PEPCK}} \text{Phospho enol pyruvate} + CO_2 + \text{GDP + Pi}$$

The enzymes, pyruvate carboxylase and PEPCK are necessary for gluconeogenesis, because the glycolytic enzyme pyruvate kinase converting phospho-enol-pyruvate to pyruvate is an irreversible enzyme. This problem is overcome by having two enzymes, namely pyruvate carboxylase and PEPCK for converting pyruvate to phospho-enol-pyruvate.

In the succeeding steps the reversible enzymes of glycolysis like enolase, phospho glycerate mutase, 1, 3-bisphosphoglycerokinase, glyceraldehyde 3-phosphate dehydrogenase, phosphotriose isomerase and aldolase are sequentially employed to generate fructose 1, 6-bisphosphate.

In the next reaction the fructose 1, 6-bisphosphate is dephosphorylated to fructose 6-phosphate by the enzyme fructose 1, 6-bisphosphatase, an irreversible enzyme which is stimulated by

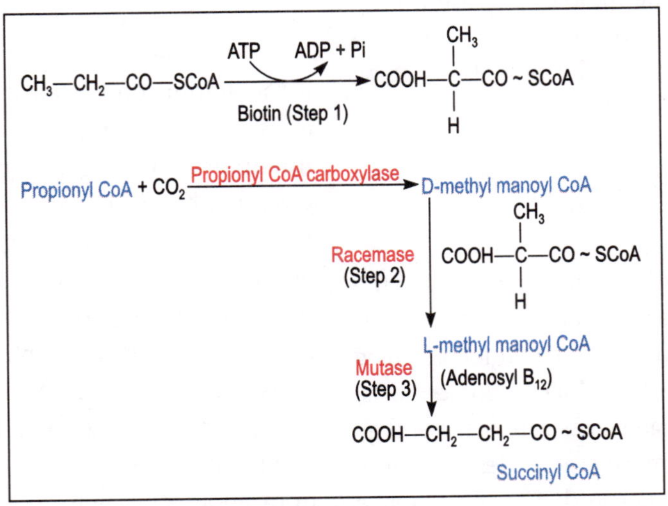

Fig. 15.6: Metabolism of propionyl CoA

hormones glucagon and glucocorticoids, but inhibited by insulin. Hence, insulin inhibits gluconeogenesis, whereas glucagon and cortisol stimulate the process. The phosphofructokinase-1, an irreversible enzyme, catalyses the opposite reaction in glycolysis.

The phosphohexose isomerase, again a reversible enzyme, hence employed in both glycolytic and gluconeogenic processes, isomerises fructose-6-phosphate to glucose-6-phosphate.

The glucose-6-phosphate is further dephosphorylated to free glucose by glucose-6-phosphatase which also serves to catalyse the final step in glycogenolytic process. This enzyme is activated by glucagon and glucocorticoids, but inhibited by insulin, in a similar way these hormones influence the first three rate limiting enzymes (pyruvate carboxylase, PEPCK and fructose 1, 6-bisphosphatase) of gluconeogenesis. Hence, the gluconeogenesis is stimulated and glycolysis is inhibited during fasting.

The gluconeogenesis gains importance when once the liver glycogen stores are depleted as a result of fasting for about 18-20 hours in an effort to restore blood glucose levels. Being an anabolic pathway, gluconeogenesis requires 4 ATP moles for the synthesis of one mole of glucose from 2 moles of pyruvate. Compounding this is a concurrent energy deficit in the form of ATP being less available during fasting.

GLYCOGEN METABOLISM

Glycogen is a homopolysaccharide present in the liver, muscle and to some extent in the kindeys. The hepatic tissue of glycogen is mainly for providing glucose units to blood for glucose homeostasis, whereas muscle glycogen takes care of the energy requirements during muscle activities. It cannot contribute to blood glucose due to the absence of glucose 6-phosphatase enzyme in muscles. Glycogen metabolism is comprised of glycogenesis and glycogenolysis.

Glycogenolysis

This is a process by which glycogen is degraded to glucose in the liver, and glucose-6-phosphate in the muscles. These reactions occur in the cytoplasm. The first enzyme in glycogenolysis is glycogen phosphorylase which hydrolyses the alpha 1-4 glycosidic bond from the periphery of glycogen to result in glucose 1-phosphate and a glycogen molecule 1 glucose unit less. The enzyme requires pyridoxal phosphate as coenzyme and inorganic phosphatase.

$$\text{Glycogen} + \text{Pi} \xleftrightarrow[\text{}]{\text{Pyridoxal phosphate / Glycogen phosphorylase}} \text{Clycogen (-1 glucose)} + \text{glucose 1 phosphate}$$

The glycogen phosphorylase attacks the alpha 1-4 glycosidic linkages sequentially till it reaches a glucose residue which is 3-4 glucose residues away from a branch point with alpha 1-6 glycosidic bond. As this enzyme is unable to act on the alpha 1-6 glycosidic linkages, there is the requirement of a bifunctional enzyme having glucan transferase and debranching enzyme activities. The glycan transferase first removes a trisaccharide unit from the branch point and transfers the three glucose residues to an adjacent linear branch of glycogen by an alpha 1-4 glycosidic bond. The exposed alpha 1, 6-glycosidic branch point, having a single glucose unit, acted upon by the debranching enzyme leads to the removal of the glucose as free glucose. After the removal of the branch point, the remaining glucose units are again sequentially cleaved off by the phosphorylase enzyme till it further reaches another glucose residue, 3-4 glucose units away from another branch point having alpha 1, 6-glycosidic linkage.

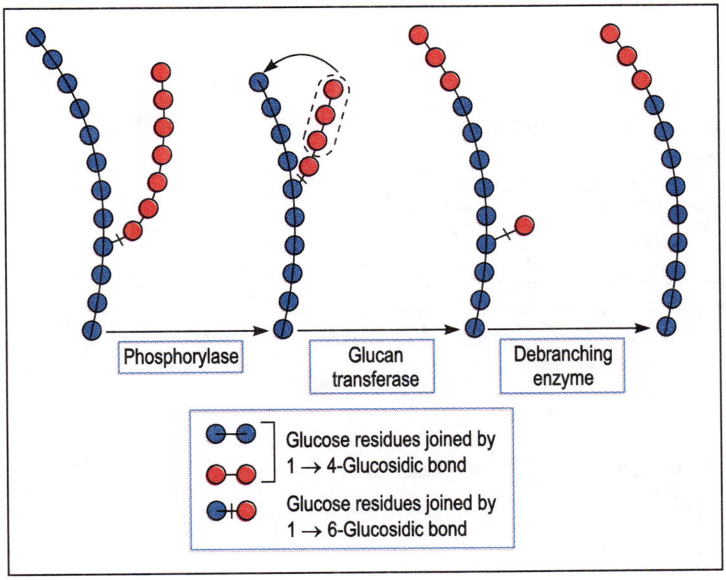

Fig. 15.7: Glycogenolysis

The action of glycogen phosphorylase, in the absence of debranching enzyme, results in one type of glycogen storage disease, the limit dextrinosis, in which a highly branched limit dextrin accumulates in the body.

The action of debranching enzyme produces free glucose, whereas, the action of glycogen phosphorylase forms glucose 1 phosphate which is subsequently converted to glucose 6-phosphate by phosphoglucomutase.

In the next step, glucose 6 phosphate forms free glucose with the help of the enzyme glucose-6-phosphatase which also is required in gluconeogenic pathway. The deficiency of this enzyme causes type 1 Glycogen storage disease.

Regulation of glycogenolysis

Epinephrine and glucagon stimulate the breakdown of glycogen in the liver, whereas insulin has the opposite effect. At the same time, the skeletal muscle phosphorylase is activated by epinephrine while glucagon has no role in regulating glycogenolysis in the muscle tissue. The calcium acting through calmodulin also stimulates muscle glycogenolysis. The glycogen synthesis and breakdown are regulated in such a manner that at a give point of time there is either synthesis or degradation of glycogen taking place. This prevents futile cycles and has resemblance to the reciprocal regulation of glycolysis and gluconeogenesis.

Hormonal regulation

The epinephrine in skeletal muscles and glucagon and epinephrine in the liver activate the enzyme adenyl cyclase situated in the cell membranes. First these hormones bind to specific receptors on the cell membrane and converts the inactive form of adenyl cyclase to the active form. The active adenyl cyclase acts on ATP and forms cyclic 3′5′ AMP liberating a pyrophosphate. The cyclic AMP acting as second messenger serves as an intracellular signal for extracellular events and subsequently activates an enzyme of broad specificity, the cyclic AMP-dependent protein kinase which has a catalytic (a) and regulator (r) subunits. The enzyme is inactive when the two

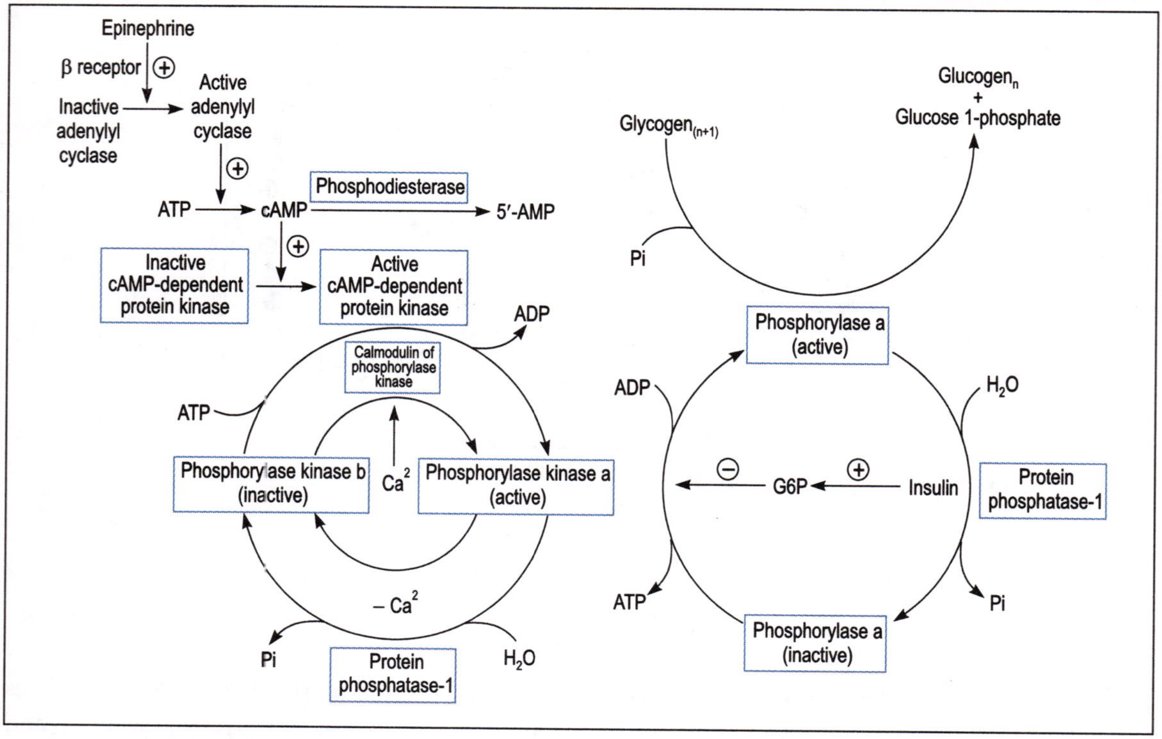

Fig. 15.8: Regulation of glycogenolysis

subunits are combined. The cyclic AMP combines with the regulator unit when the catalytic unit becomes free which in turn activates an inactive phosphorylase kinase by phosphorylating the hydroxyl groups of its amino acids, serine and threonine. This phosphorylation requires ATP. The enzyme phosphorylase kinase has 4 subunits, namely, alpha, beta, gamma and delta. The gamma is the catalytic unit while the others are regulatory in nature. Phosphorylation and binding with calcium makes the enzyme active. The delta unit has resemblance to the calcium binding protein calmodulin. It is to be emphasised that calcium released from sarcoplasmic reticulum stimulates both muscle contraction and glycogenolysis in muscle.

The phosphorylase kinase, in turn, activates an inactive glycogen phosphorylase, again by phosphorylation of the hydroxyl groups of serine and threonine of the enzyme. The key enzyme of glycogenolysis, the active glycogen phosphorylase, hydrolyses glycogen as mentioned earlier. The cyclic AMP can activate 2 additional enzymes: phosphorylase kinase and glycogen phosphorylase but it inhibits glycogen synthase. Hence, addition of phosphate makes the glycogenolytic enzymes active, whereas adding phosphate to the key enzyme of glycogenesis, makes it inactive.

Hence, epinephrine and glucagen, secreted in response to hypoglycaemia, brings abut the degradation of glycogen to glucose and inhibits glycogenesis. In contrast insulin, secreted in response to hyperglycaemia inhibits glycogenolysis but stimulates glycolysis.

Cyclic AMP, after its function, is degraded to AMP by phosphodiesterase which is stimulated by insulin, thereby decreasing the cyclic AMP levels. Caffeine, theophylline and theobromine inhibit phosphodiesterase, resulting in increased glycogenolysis and also lipolysis, the latter leading

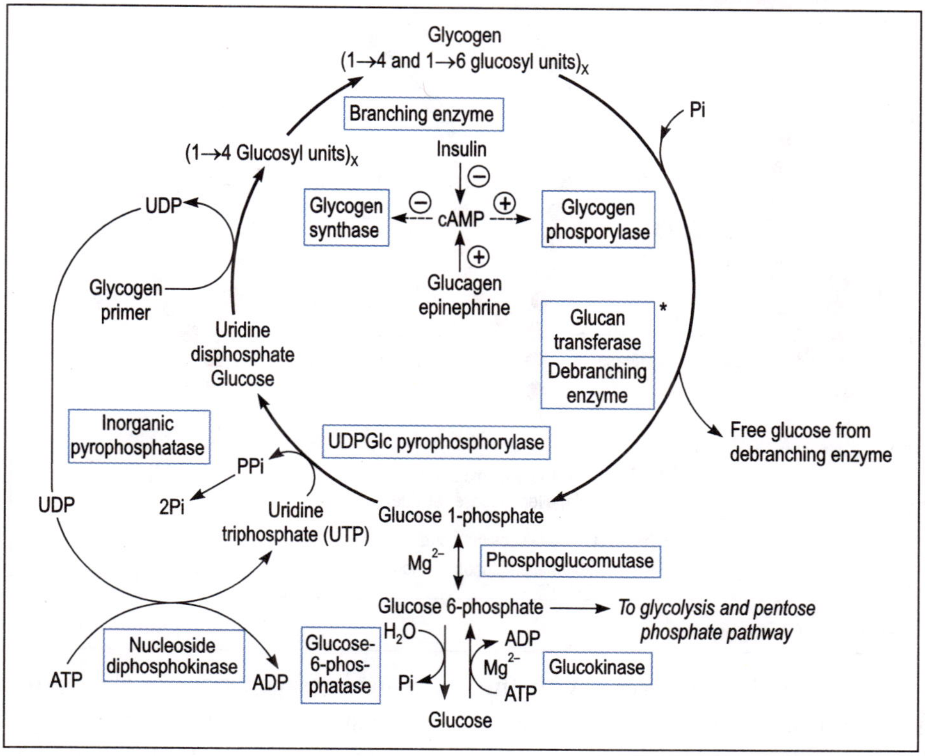

Fig. 15.9: Glycogen metabolism

to hypercholesterolaemia from acetyl CoA obtained from the beta oxidation of fatty acids. This could explain the deleterious effects of excessive intake of tea and coffee.

Glycogenesis

Glycogenesis is the process of formation of glycogen from glucose and takes place in the cytoplasm. The key enzyme of this process is glycogen synthase. First glucose is converted to glucose-6-phosphate by glucokinase in the presence of ATP and Mg^{2++}. In the next stop, glucose-6-phosphate is transformed into glucose-1-phosphate with the help of phosphoglucomutase.

1. The first committed step in glycogenesis is the reaction between glucose-1-phosphate and UTP in presence of UDP glucose pyrophosphorylase to form UDP glucose and Pi.

$$\text{Glucose 6-phosphate + UDP} \xrightarrow{\text{UDP glucose pyrophosphorylase}} \text{UDP glucose + Pi}$$

2. In the next step, the glucose from UDP glucose is transferred to a pre-existing glycogen primer molecule, glucogenin, a protein-carbohydrate complex containing an oligosaccharide of 7 glucose residues linked by alpha 1, 4 glycosidic bonds. The sequential addition of glucose units from UDP glucose molecules to the growing chain of glycogen primer is carried out by the key enzyme, glycogen synthase.

$$\text{Glycogen primer (n) + UDP glucose} \xrightarrow{\text{Glycogen synthase}} \text{Glycogen (n+1) + UDP}$$

3. As the glycogen synthase can add glucose units only by alpha 1-4 glycosidic bonds, a branching enzyme is necessary to create branch points with glucose being added by

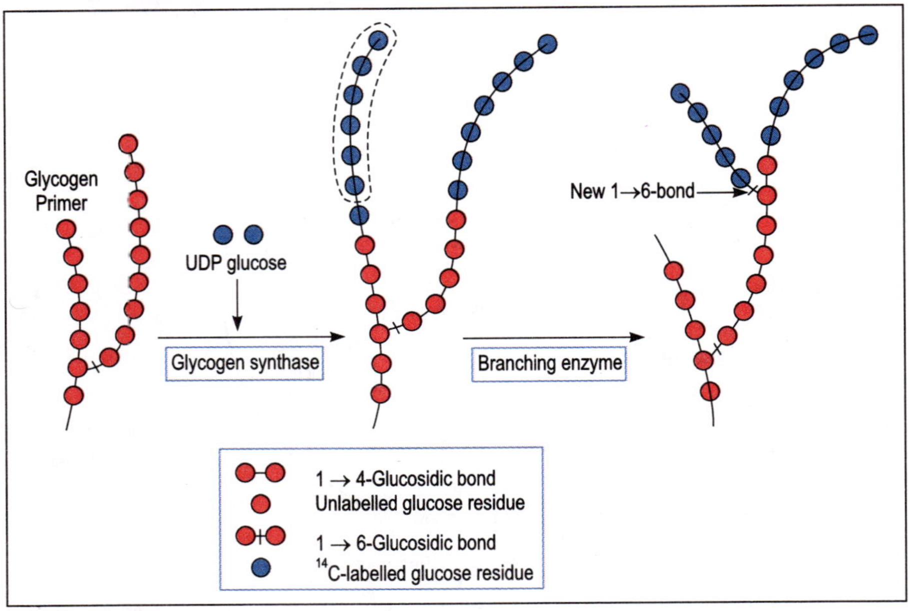

Fig. 15.10: Glycogenesis

alpha 1-6 glycosidic linkages. With the glycogen chain attaining a length of about 12 glucose residues the branching enzyme shifts an oligosaccharide of 8 glucose units from this chain to another position on the growing chain by forming an alpha 1-6 glycosidic bond.

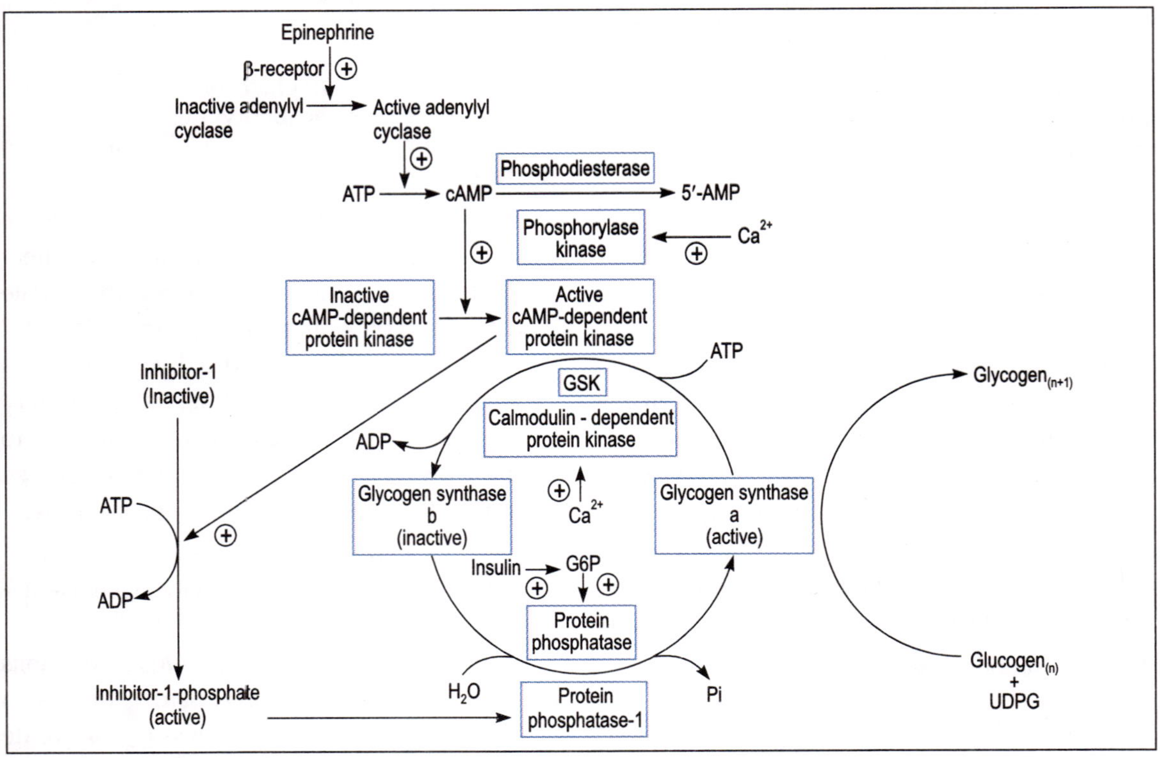

Fig. 15.11: Regulation of glycogenesis

Regulation of glycogenesis

As mentioned earlier, the key or rate limiting enzyme of glycogen synthesis is glycogen synthase which is active in a dephosphorylated state and inactive in phosphorylated state. The inactive (phosphorylated) glycogen synthase is activated by protein phosphatase 1, which is activated by insulin hormone. The protein phosphatase 1 removes phosphate from glycogen synthase, thereby making it active. In contrast, the cyclic AMP dependent protein kinase, formed in response to the activation of adenyl cyclase, can phosphorylate glycogen synthase making it inactive. As pointed out earlier, the phosphorylation of phosphorylase kinase, glycogen phosphorylase and synthase kinase by cyclic AMP dependent protein kinase makes these enzymes active. This is facilitated by glucagon and epinephrine. Further, the protein phosphatase 1, which activates glycogen synthase is inhibited by inhibitor 1. Hence, cyclic AMP is instrumental in regulating glycogenolysis and glycogenesis, the former being activated and the latter inhibited. This is known as the reciprocal regulation of the two processes, thereby avoiding futile cycles. The fasting leads to glycogenolysis, and glycogenesis prevails in a well-fed condition. Stated otherwise, insulin favours glycogenesis but epinephrine and glucagon brings about glycogen degradation.

Glycogen storage diseases

These are conditions in which abnormal amounts of glycogen or abnormal type of glycogen are stored in liver and skeletal muscles but this excess glycogen cannot be utilized by the body due to the absence of specific enzymes in the degradation or the synthesis of glycogen. These are classified as inborn errors of metabolism. Several types of glycogen storage diseases are known:

Type 1 glycogen storage disease

This is also termed as von Gierke's disease. This is due to the deficiency of the enzyme, glucose-6-phosphatase, which is required for the generation of glucose from glucose-6-phosphate by removing the phosphate. As the liver glycogen cannot be degraded to glucose due to want of the last enzyme (glucose-6-phosphatase) of glycogenolysis. Only free glucose is permeable through the cell membrane of hepatocytes. Hence, in type 1 glycogen storage disease, liver glycogen does not provide glucose units to the blood, resulting in hypoglycaemia specially while fasting. Ketosis may result due to the excessive oxidation of fatty acids and is reinforced by the less availability of oxaloacetate (which gets routed for gluconeogenesis) for condensing with acetyl CoA to form citrate. The acetyl CoA is forced to take the alternative pathway of ketogenesis leading to ketosis.

Further, the accumulated glucose-6-phosphate can be going to the HMP shunt pathway, thereby giving rise to excess of ribose phosphate which could be forming more purine nucleotides, a major share of which forms uric acid, resulting in hyperuricaemia and secondary gout.

The large excess of liver glycogen can cause cirrhosis. The condition is fatal. Food at frequent intervals may extend the life-span of the patient by a few months.

The Type 1 disease is more commonly encountered than the other types.

The Type 2 disease is known as Pompe's disease and is due to the absence of lysosomal maltase which culminates in the excess deposition of glycogen in liver and muscles.

The type 3 is as a result of the absence of the debranching enzyme in which case there is failure to break alpha 1, 6 glycosidic bonds of the highly branched glycogen molecule resulting in the accumulation of highly branched dextrin. Hypoglycaemia during fasting is present. This condition is referred to as limit dextrinosis.

The amylopectinosis or Anderson's disease (type 4) has deficient branching enzyme, leading to the deposition of glycogen with very less branch points. The branch points are formed by alpha 1, 6 glycosidic linkages by the combined action of glucan transferase and a branching enzyme. Hypoglycaemia is often present.

The patients with McArdle's disease (type 5) has a decreased tolerance for exercise because the muscles lack glycogen phosphorylase enzyme. The muscle, though rich in glycogen, cannot put it to use for energy production.

Other types of glycogen storage disorders reported include those due to the deficiency of any of the enzymes like liver phosphorylase, muscle phosphofructokinase, liver phosphorylase kinase and muscle phosphorylase kinase.

Hexose monophosphate pathway (Pentose phosphate pathway)

It is also known as phosphogluconate pathway. As pentoses are involved, it can further be termed as pentose phosphate pathway. The HMP shunt pathway is very active in erythrocytes, liver, lactating mammary glands, adrenal cortex, testes, ovaries, adipose tissue and lens of the eye. In addition to producing ribose phosphate, for the nucleotide synthesis, the HMP shunt serves as the major metabolic pathway enabling the body to produce NADPH, essential for the synthesis of cholesterol, bile acids, fatty acids and steroid hormones. The functions of NADPH also comprises of its antioxidant-free radical scavenging properties like conversion of oxidised to reduced glutathione, reduction of methemoglobin to haemoglobin, etc. Two NADPH moles are produced from every glucose molecule entering the oxidative phase of the pathway and there is liberation of CO_2 from the 1st carbon of glucose 6 phosphate.

The HMP shunt pathway has oxidative and non-oxidative phases. This takes place in the cytosol. The oxidative part of the pathway, comprising 3 reactions, produces ribulose 5 phosphate, 2 molecules of NADPH and one mole of CO_2. This phase is quite significant in those tissues requiring NADPH for reductive biosynthesis of fatty acids, like lactating mammary glands, liver and adipose tissue. Other tissues like testes, ovaries, placenta and adrenal cortex require NADPH for the synthesis of steroid hormones. The synthesis of bile acids in the liver needs a NADPH.

Oxidative phase

Step 1: The enzyme glucose-6-phosphate dehydrogenase (G6PD) irreversibly oxidises glucose-6-phosphate to 6-phospho-gluconolactone. The hydrogens are accepted by NADP to get reduced to NADPH. The G6PD is the regulatory enzyme of the pathway, with NADPH an inhibitor of the enzyme. Insulin favours the G6PD gene expression and hence, the pathway increases in a well-fed state.

Step 2: The 6-phosphogluconolactone is converted to 6-phosphogluconate by 6-phopho-gluconolactone hydrolase by the addition of water.

Step 3: The 6-phosphogluconate is oxidatively decarboxylated to ribulose-5-phosphate with the removal of CO_2 and the accompanying reduction of NADP to NADPH. This is the 2nd molecule of NADPH generated in HMP shunt pathway. All the above reactions are irreversible. The CO_2 removed is from 1st carbon of glucose.

Nonoxidative phase

These reactions take place in all the cells and the ribose phosphate formed is used for the formation of nucleotides and nucleic acids. This phase functions to interconvert sugars having 3-7 carbons. All these reactions are reversible in nature.

Step 4: The ribulose-5-phosphate can either be isomerised to ribose-5-phosphate by

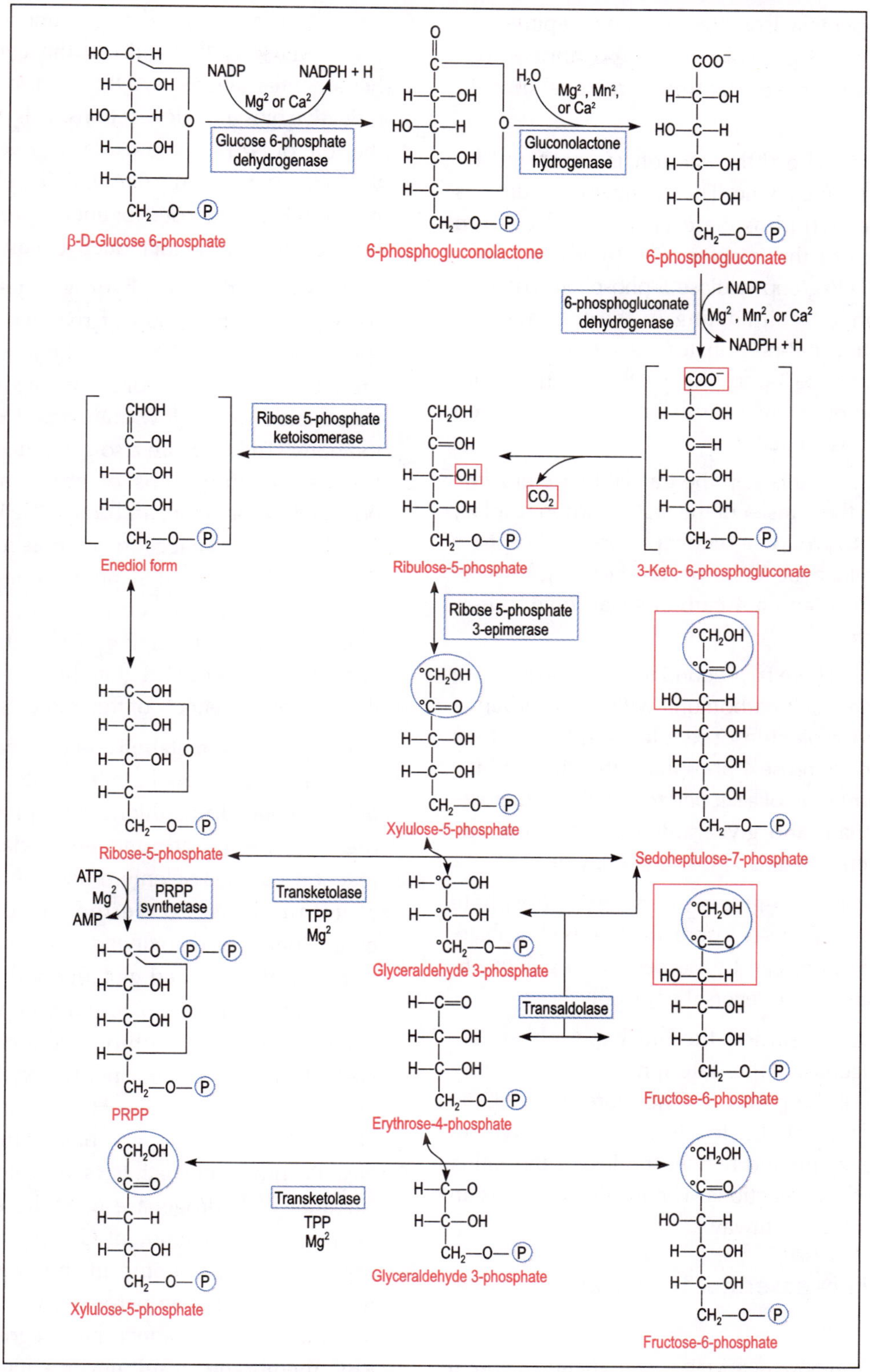

Fig. 15.12: Hexose monophosphate shunt pathway

phosphopentose isomerase or can be epimerised to xylulose-5-phosphate by phosphopentose epimerase. These reactions are reversible in nature.

Step 5: The fifth reaction is catalyzed by transketolase enzyme. The coenzyme required is thiamine pyrophosphate (TPP). The reaction comprises of the transfer of a 2-carbon moiety having a keto group xylulose-5-phosphate to ribose-5-phosphate, culminating in the formation of sedoheptulose-7-phosphate (a 7-carbon sugar) and glyceraldehyde 3-phosphate, a three-carbon unit. Thiamine deficiency can have adverse effect on transketolase reactions.

Step 6: In this step the transaldolase enzyme catalyses the transfer of the first 3-carbon residue, with a keto group, of sedoheptulose-7-phosphate to glyceraldehyde 3-phosphate generating fructose-6-phosphate and a 4-carbon sugar, erythrose-4-phosphate.

Step 7: There is a second transketolase step in which the first 2-carbon unit with a keto group of a second molecule of xylulose-5-phosphate is shifted to erythrose 4-phosphate, thereby resulting in the formation of a second molecule of fructose-6-phosphate and glyceraldehyde 3-phosphate. This reaction needs TPP as a coenzyme.

Step 8: Two-glyceraldehyde 3-phosphate molecules condenses to form fructose-6-phosphate, which, in turn, can be isomerised to glucose-6-phosphate by the reversal of glycolysis.

As stated earlier, the glucose-6-phosphate dehydrogenase is the rate limiting or regulatory enzyme of the pathway. The more the NADPH status of the cell, the less is the rate of activity of G6PD enzyme and hence, the less is the influx through these reactions. The insulin serves as an inducer of the pathway.

NADPH is essential for several processes

1. The structure of NADP is only slightly different from NAD, in having an additional phosphate group, but has entirely different functions. The low NADP/NADPH ratio in the cytosol of hepatocytes allows NADPH to function in reductive biosynthetic reactions. This contrasts the role of NAD in the oxidative role due to the high ratio of NAD^+/NADH. The electrons of NADPH are not used for energy production but are rather meant for reductive biosynthesis.
2. It is used to detoxify hydrogen peroxide, belonging to the group of reactive oxygen species (ROS). The ROS are continuously being produced during metabolic reactions and may be coupled with a low antioxidant status in the body. The ROS are also generated during the detoxifying steps of environmental pollutants, xenobiotics and drugs. The ROS can cause severe damages to biomolecules like PUFA, cell membranes, proteins and DNA. This can lead to cell death, cancer atherosclerosis and aging. The antioxidant enzymes, vitamins and glutathione minimise the deleterious effects of free radicals.
3. The erythrocyte metabolism generates H_2O_2 which is detoxified to harmless water with the help of reduced glutathione in the presence of the enzyme glutathione peroxidase. The reaction requires selenium and the reduced glutathione is converted to oxidised glutathione. For detoxifying more molecules of H_2O_2, the reduced glutathione has to be regenerated from the oxidised from which is carried out by glutathione reductase, an NADPH dependent enzyme. The erythrocytes entirely depends on the HMP shunt to get NADPH as they do not have alternative NADPH producing pathways like cytoplasmic isocitrate dehydrogenase or malic enzyme. Therefore, a deficiency of G6PD enzyme in erythrocytes, can come in the way of the removal of hydrogen peroxide, which when accumulated, may affect the integrity of the RBC membrane, resulting in a haemolytic condition. This could subsequently lead to

unconjugated hyperbilirubinaemia (haemolytic jaundice). The G6PD deficiency decreases the formation of NADPH and hence, care should be taken while prescribing oxidant type of drugs like primaquin, aspirin, sulpha drugs, etc. Hence, malarial patients with G6PD deficiency may suffer from haemolytic crisis, if primaquin is administered, as the demand for NADPH increases in this condition. The G6PD deficiency is an x-linked condition and is one of the most commonly detected enzyme deficiencies, mainly seen in Asian, African and the middle east populations. At the same time it is noteworthy that G6PD deficiency gives resistance to plasmodium falciparum infection. The other enzymes involved in the conversion of reactive oxygen species include super oxide dismutase and catalase which along with G6PD form a group of antioxidant enzymes serving to defend against the bad effects of the highly reactive free radicals.

4. There are antioxidant vitamins like vitamin E, ascorbic acid and beta carotene which can detoxify ROS. Hence, these vitamins may have anticancer effect. The beneficial effects of fruits and vegetables probably stem from the fact that these foods are rich in antioxidants.

5. In the cytochrome P450 mono-oxygenase system, it is the NADPH that provides the reducing equivalents for the metabolism of steroids, drugs and xenobiotics. The mitochondrial cytochrome P450 along with NADPH is involved in the steroid formation in the placenta, testes, ovaries and adrenal cortex from cholesterol. The activation of cholecalciferol in the liver and kidneys to form calcitriol (1, 25 dihydroxy cholecalciferol) also requires the same enzyme system involving NADPH. NADPH converts met-hemoglobin to hemoglobin.

6. The microsomal NADPH dependent cytochrome P450 mono-oxygenase system of hepatic smooth endoplasmic reticulum is very important in the detoxification of several drugs and foreign compounds. The drugs may be made more active or inactive. The toxins are made water soluble so that these can be excreted from the body. Or else, it may be conjugated with compounds like glucuronic acid before excretion.

7. Phagocytosis of micro-organisms and foreign particles by neutrophils and monocytes is an important defence mechanism. The NADPH oxidase and myeloperoxidase present in macrophages degrade micro-organisms. The combined action of the above two enzymes is bactericidal. The NADPH oxidase produces superoxide anion and this process is termed as respiratory burst. The deficiency of NADPH oxidase causes chronic granulomatous disease characterised by infections. The superoxide is further converted to H_2O_2 by superoxide dismutase. Next, the myeloperoxidase, a haem containing enzyme converts H_2O_2 to hypochlorous acid in presence of chloride. The hypochlorous acid is bacteroidal in nature.

8. Yet another function of NADPH is in the formation of nitric oxide from arginine. The cytosolic nitric oxide synthase forms nitric oxide and citrulline in the presence of arginine, oxygen, NADPH, FMN, FAD, haem and tetrahydrobiopterine. Two of the three NO syntheses are formed at constant rate and are dependent on calcium-calmodulin system. These enzymes, eNOS and nNOS are seen in the endothelium and the nervous tissues respectively. The third enzyme is an inducible one and calcium dependent and is present in all cells. The inducers include bacterial toxins, cytokines, etc.

The drugs, nitroglycerine and nitroprusside, are also converted to nitric oxide which causes vascular smooth muscle relaxation, thereby reducing blood pressure. The NO formed by eNOS or from the above mentioned drugs, activates guanylate cyclase forming cyclic GMP which, in turn activates, cyclic GMP dependent protein kinase. The protein kinase causes decreased entry of calcium into the smooth muscle cells, by inactivating calcium channel proteins, achieved by phosphorylating these proteins. The decreased entry of calcium into smooth muscle cells invariably results in its relaxation and the accompanying decrease in blood pressure. This can further be explained by a decreased calcium-calmodulin activation of myosin light chain kinase, causing inhibition to smooth muscle contraction.

The NO synthesis is greatly increased in bacterial infection, which causes the induction of macrophage NO synthase. The macrophages also generate the ROS like superoxide which along with NO can subsequently cause the synthesis of other highly reactive free radicals like hydroxyl radical, detrimental to micro-organisms.

Further, NO prevents the platelet aggregation and also acts as a neurotransmitter. The DNA sequencing of G6PD gene has revealed as many as 400 mutations of the gene and life will not be compatible in case of complete absence of the enzyme, as the patient could suffer from severe haemolysis.

Blood glucose regulation

The fasting blood glucose level is required to be maintained between 70 to 110mg/dl. This is accomplished by a variety of factors like the nutritive state and the actions of hormones like insulin, on one hand, and glucagon, glucocorticoids, epinephrine and growth hormone on the other, making it a finely regulated process. The insulin is the only hypoglycaemic hormone, all the others being hyperglycaemic in nature.

Insulin is a polypeptide hormone, synthesised and secreted by the beta cells of islets of Langerhans of pancreas in response to elevated levels of glucose and amino acids in the blood.

Insulin synthesis and secretion are decreased during fasting, fever and infection. This is mainly mediated by epinephrine liberated by sympathetic nervous system in response to stressful situations and this leads to the release of glucose obtained from glycogenolysis and gluconeogenesis. The epinephrine also brings about adipose lipolysis and release of fatty acids.

The insulin is synthesised as pre- and pro-insulin, which has a total of 85 amino acids, consisting of signal sequence at its amino terminus, which is present in almost all the secretory proteins for entering the endoplasmic reticulum. After entering the ER the signal peptide is cleaved off forming pro-insulin having a connecting peptide (C-peptide). The pro-insulin is further activated to insulin by the removal of the connecting peptide, leaving 2-peptide chains, and an A-chain with 21 amino acids and a B-chain with 30 amino acids, the two chains being linked by 2 inter-chain disulfide bonds cysteine residues, namely, one disulfide bond between the 7th cysteine of A-chain and the 7th cysteine of B-chain, and the second, between the 20th cysteine of A-chain and 19th cysteine of B-chain. Additionally, an intra-chain disulfide bond exists between 6th and 11th cystine residues of A-chain. The human insulin differs from the insulin of other animals like ox in the amino acids at numbers 8, 9 and 10 of the A-chain. The porcine insulin is least different from human insulin in the sequence of amino acids, with the sole difference of alanine replacing threonine at the 30th position on the B-chain of porcine insulin. Hence, de-alaninated porcine insulin is resembling human insulin. The use of insulin from other animals for the treatment of diabetes mellitus may form antibodies in humans. With the availability of insulin by DNA recombinant technology, the problem of antibody production has been circumvented. The stored insulin is released into the blood primarily in response to hyperglycaemia.

Functions of insulin

Insulin increases the number of Glut-4 receptors in muscle and adipose tissue with the resultant

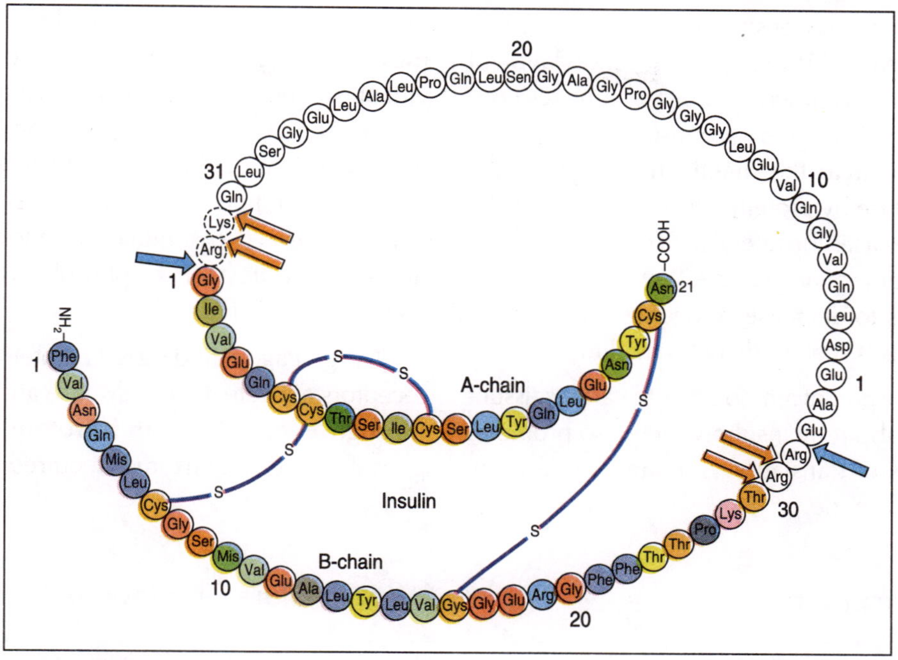

Fig. 15.13: Insulin molecule with connecting peptide

increased uptake of glucose by these tissues leading to the lowering of blood glucose level. Insulin stimulates glycogenesis in muscle and liver. It also decreases glycogenolysis and gluconeogenesis in liver. Further, glycolysis is activated by insulin. Glucagon, glucocorticoids and epinephrine increase blood glucose level, whereas insulin decreases the blood sugar concentration.

Insulin is a lipogenic and antilipolytic hormone, meaning it favours lipid synthesis and prevents triglyceride breakdown in adipose tissue by

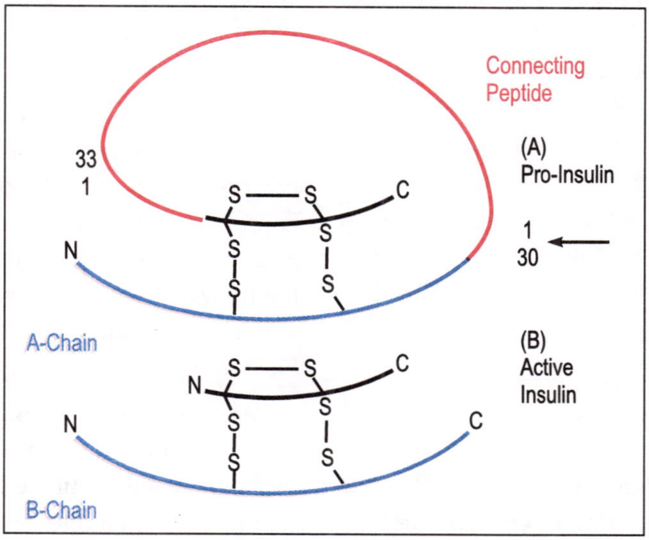

Fig. 15.14: Conversion of pro-insulin to active insulin. Arrows = site of action of proteolytic enzymes

inhibiting hormone sensitive lipase. The inhibition of the lipase is achieved by dephosphorylation of the phosphate from serine and threonine residues of the enzyme lipase by a phosphatase enzyme. This is similar to the inactivation of glycogen phosphorylase by protein phosphatase-1. Insulin further decreases the rate of beta oxidation and so the generation of acetyl CoA. Yet another action of insulin is to decrease ketogenesis and hence, ketoacidosis is seen in diabetes mellitus.

The transport of amino acids to various tissues is brought about by insulin which also promotes the process of translation, serving its role as an anabolic hormone.

Insulin receptors

The insulin receptor consists of two alpha and two beta subunits aggregating into a tetramer. The alpha subunits are outside the membrane and has insulin binding site. The beta subunits are transmembrane proteins and projecting into the cytoplasmic side. The binding of insulin to alpha subunits causes conformational changes on the beta subunits which possesses tyrosine kinase activity which phosphorylates insulin receptor substrates. This starts the activation of several processes like gene expression and metabolism. Insulin favours the recruitment of Glut-4 receptors in adipose tissue and muscle so that glucose transport into these tissue is facilitated. Insulin induces the synthesis of glucokinase, phosphofructokinase, pyruvate kinase, acetyl CoA, carboxylase, fatty acid synthase and glycogen synthase.

Functions of glucagon

The glucagon is a peptide hormone with 29 amino acids secreted by the alpha cells of islets of Langerhans of pancreas in response to hypoglycaemia. Glucagon stimulates glycogenolysis and gluconeogenesis in liver but not in muscle. The cortisol or glucocorticoids also cause gluconeogenesis and hence, is hyperglycaemic in nature. The glucagon also brings about lipolysis in adipose tissue. Epinephrine and norepinephrine also stimulate lipolysis. Therefore, glucagon and epinephrine act to stimulate hormone sensitive lipase, via cyclic AMP, by phosphorylating lipase and causing lipolysis.

The glucagon binding to G protein dependent receptors activate adenyl cyclase and further act through cyclic AMP and protein kinases to phosphorylate the enzymes required for lipolysis and glycogenolysis.

Abnormalities in blood glucose

Hypoglycaemia is when the glucose level reaches about 40 to 45mg/dl. As the brain has an absolute need for glucose as energy fuel, hypoglycaemia leads to confusion, coma and death if not attended to on an emergency basis. Prolonged hypoglycaemia is life threatening and the body has mechanisms to defend itself like low insulin levels and elevated concentration of glucagon, epinephrin and cortisol. Hypoglycaemic patients suffer from brain dysfunction, seizures, slurred speech, anxiety, etc.

Hypoglycaemia can occur in diabetic patients on insulin treatment, when the dose of insulin is not properly standardised. This can occur in a conscious or in an unconscious state. If the patient is conscious he can be given glucose with which the normal level of glucose can be restored. The unconscious patients with hypoglycaemia may have to be administered glucagon.

Another type of hypoglycaemia is due to an excessive release of insulin in response to the ingestion of meal, but the glucose level, however, returns to normal after some time. Such patients can be advised to take small frequent metals rather large meals.

A third type of hypoglycaemia is encountered in patients with hepatic damage leading to decreased glycogenolysis and gluconeogenesis during fasting. This may be seen in people with pancreatic alpha cell dysfunction and also in adrenal insufficiency.

An insulin producing alpha cell tumour of pancreas can cause increased synthesis and release of insulin which causes increased peripheral utilisation of glucose resulting fasting hypoglycaemia.

The intake of alcohol may result in hypoglycaemia. Alcohol is first acted upon by alcohol dehydrogenase, in presence of NAD to acetaldehyde and NADH + H^+. The acetaldehyde formed is further dehydrogenated to acetic acid by the enzyme acetaldehyde dehydrogenase, a second NAD being converted to NADH + H^+. The elevated levels of NADH converts the gluconeogenic substrates, pyruvate and oxaloacetate to lactate and malate respectively, which leads to reduced gluconeogenesis and resultant hypoglycaemia. There is a possibility of acetyl CoA being diverted to ketogenic pathway. The symptoms (behavioural changes) of alcoholism will be exaggerated, when the person is habituated to take less food, as is definitely with alcoholics. As the food intake is decreased in the alcoholics there will be less glycogen stores in the liver. Further, chronic alcoholism can cause fatty liver as a fallout of decreased fatty acid oxidation due to a fall in NAD/NADH ratio. This condition leads to the formation TAG from fatty acids and glyceraldehyde 3-phosphate, a process known as lipogenesis. These triglycerides are deposited in the liver because all the TAGs could not be packaged into VLDL and secreted into the blood by the liver. The fatty liver, later on, develops into hepatitis and liver cirrhosis.

Diabetes Mellitus

Diabetes mellitus is a multifactorial disease in which the blood glucose is elevated above the normal level of 70-110mg/dl. This is caused by an absolute lack of insulin or due to peripheral resistance to the action of insulin as a result of deficient or defective insulin receptors. Diabetes of long-term duration may cause renal insufficiency, retinopathy, angiopathy, stroke and heart attack. The diabetes mellitus can be classified into two types:

Type 1, also called as insulin dependent and type 2, also called as non-insulin dependent. Type 1 and Type 2 classification is more acceptable.

Type 1 Diabetes Mellitus

It occurs during childhood or below 30 years of age and constitutes about 10 out of every 100 cases diagnosed. There is complete destruction of the beta cells due to an auto immune condition. A virus infection can also cause beta cell destruction. The patients are mostly undernourished and are thin. The plasma insulin levels are low or absent, ketoacidosis is common and there is no response to oral hypoglycaemic medications. The treatment requires insulin administration. There is about 10 per cent genetic predisposition for type 1 DM.

The symptoms of this type of diabetes include polyphagia (increased hunger), polyuria (increased urine output) and polydypsia (increased thirst), along with weight loss and weakness. The diagnosis of diabetes is done if the FBS is more than 126mg/dl at least on two occasions. There could be accompanying ketoacidosis.

Metabolic alterations

In DM, there is increased gluconeogenesis and decreased glycolysis due to a deficiency of insulin and an excess of glucagon and cortisol. This results in hyperglycaemia. The Glut-4 receptors, sensitive to insulin, is present in muscle and adipose tissue and these receptors become dysfunctional in insulin lack. The diabetic ketoacidosis results because of the activation of hormone sensitive lipase that leads to the breakdown to TAG in adipose tissue and the release of fatty acids into plasma. These NEFA are transported with albumin to various tissues which take up these fatty acids. As the glucose is not utilised by the cells, fatty acids undergo beta oxidation to form acetyl CoA. The excessive amounts of acetyl CoA cannot enter the TCA cycle due to want to oxaloacetate, as oxaloacetate is diverted for the synthesis of glucose although glucose is unusable in DM patients. Hence, the acetyl CoA goes for the formation of the ketone bodies like acetoacetate, beta hydroxy butyrate and acetone, which are all acidic. This cause diabetic ketoacidosis (DKA), if not treated with insulin, fluids and electrolytes. In DKA the body tries to use the buffers like bicarbonate, phosphate, proteins and haemoglobin in order to neutralize these ketone bodies. The sodium bicarbonate reacts with acetoacetate to form sodium acetoacetate and carbonic acid. This process converts a strong acetoacetic acid to a weak acid, so that the pH change remains minimal. The beta hydroxy butyrate is also neutralized similarly. But the buffers fails to cope up with the excessive production of ketone bodies. The kidneys excrete acidic urine in an endeavour to maintain the pH. The lungs excrete more CO_2 by hyperventilation. Further, the acids are excreted as salts of sodium and potassium by the kidneys causing alkali deficit in the body leading to a state of ketoacidosis.

The administration of insulin reverses the whole process. The insulin corrects hyperglycaemia by removing glucose from the blood to the adipose tissue and also to muscles, by influencing Glut-4 receptors. Within the cell, insulin stimulates glycolysis and glycogenesis but depresses gluconeogenesis and glycogenolysis. The insulin acting as an antilipolytic hormone inhibits hormone sensitive lipase and decrease adipose tissue lipolysis. This leads to a decrease in plasma fatty acids which in turn leads to decreased uptake of fatty acids by tissues. Subsequently, the oxidation of fatty acids is reduced, so is the acetyl CoA synthesis and ketogenesis. The acetyl CoA, now can combine with oxaloacetate and start the TCA cycle. With the availability of insulin, glycolysis commences, resulting in pyruvate which can form oxaloacetate. With the cessation of ketogenesis, ketosis is corrected and the blood pH is normalised.

A very important test done to know as to how much the treatment of diabetes mellitus is effective, is the estimation of glycosylated haemoglobin, HbA1C, the normal value is around 6 per cent or less of the total haemoglobin. The glyco-Hb level of a diabetic patient, measured, today is indicative of the glucose level, about three months before and it bears no association with the glucose level of the patient, measured today. This is important as the complications of diabetes, like retinopathy, nephropathy and neuropathy are due to an elevated level of plasma glucose.

Type 1 diabetic patients are also prone to slip into hypoglycaemia as these patients may develop glucagon deficit. They also may fail to secrete epinephrine, a few years after detection of diabetes, thereby causing episodes of hypoglycaemia which in children can hamper brain development.

Type 2 Diabetes Mellitus

It constitutes about 90 per cent of the diabetic cases. Here the insulin levels are normal or elevated but the cardinal feature is resistance to insulin action. The obesity increases insulin resistance, whereas the insulin resistance is decreased with loss of body weight. Further, the elevated levels of plasma fatty acids, encountered in diabetes, due to lipolysis of adipose tissue TAG can also cause insulin resistance. It can be the outcome of defective insulin receptors also. Some type 2 patients may have a superimposed defective beta cell function also. The type 2 diabetes develops after 35 years of age and exhibits polyuria, polydypsia and polyphagia. Initially the patients are obese or overweight and are inactive, and later suffers from weight loss. The type 2 patients can be managed with oral hypoglycaemic agents in the initial stages but may need insulin in the later stages. They are less prone to develop DKA as the insulin present in small quantities can serve to overcome ketosis, but may have hyperosmolar nonketotic comma.

The long term complications of type 2 diabetes include nephropathy, retinopathy, neuropathy and angiopathy. The nephropathy is due to the glycation of the basement membrane proteins of renal tissue causing renal damage. The retinopathy is because of the generation of sorbitol from glucose which accumulates in the lens of the eyes and the sorbitol being an osmotically active substance draws water into the lens causing swelling leading to retinopathy and cataract formation. The dulcitol also can cause cataract. The neuropathy also results from glycation of proteins of the neurons and also due to the deleterious effect of advanced glycation products. The ratio of myoinositol/inositol gets reversed. The sorbitol also is implicated in peripheral neuropathy.

The angiopathy is because of the high cholesterol levels, more specifically of high LDL and low HDL concentrations in the blood which favours the deposition of cholesterol in the subintimal layer of the walls of arteries gradually forming atherosclerotic plaques, thereby obliterating the lumen of the blood vessels. The block in the coronary arteries cause myocardial infarction and block in the cerebral vessels results in stroke. The cholesterol levels in diabetic patients are always elevated. This is due to more acetyl CoA being diverted to cholesterol synthesis as it cannot enter the TCA cycle because of a deficiency of oxaloacetate, as a result of insulin resistance.

The oxidised and glycosylated LDL, found in diabetic patients, are more atherogenic as they are taken up by macrophages in an unregulated way which is not subjected to feedback regulation unlike the LDL receptor- mediated uptake of LDL cholesterol which can be regulated by up or down regulation of the LDL receptors. The macrophages accumulate more cholesterol becoming foam cells which pass into the sub-intimal layer of the walls of the blood vessels and get deposited there to form an atherosclerotic plaque which can lead to CAD and stroke.

The diabetic patient should undergo the following tests: (a) Fasting and postprandial blood glucose to be assessed every 3 months (b) Complete lipid profile consisting of total cholesterol, HDL cholesterol, LDL cholesterol and TAG-every 6 months (c) blood urea and serum creatinine twice in a year (d) microalbuminuria to be tested once a year. The microalbuminuria is the excretion of albumin in small amounts (40 to 300mg/day) and it is a predictor of future renal damage. Another test that the diabetics should do is glyco-Hb known as HbA1C which indicates how well the glucose levels are maintained in the patient. As stated earlier, the glyco Hb level reflects the blood levels 10-12 weeks before, as the life-span of erythrocytes is 120 days.

Initially the type 2 patients have normal or a higher insulin status, but with the progression of the disease over a period of time, they develop beta cell dysfunction and would have low levels of insulin. Weight reduction, regular exercise and dietary modifications can help the patient in the early stages. As stated earlier, the type 2 patient will be needing oral hypoglycaemic drugs and insulin in the later years.

Some of the pregnant ladies may have gestational diabetes which can further develop into diabetes mellitus type 2. Some other males and females can be suffering from impaired glucose tolerance which also can develop to type 2 diabetes mellitus.

Metabolic syndrome

This is a disorder in which at least any 3 of the following conditions persist: abdominal obesity, glucose intolerance, insulin resistance, hyperinsulinaemia, dyslipidaemias, (namely, elevated LDL and TAG, but lower HDL) and high blood pressure. It has been observed that the low levels of the adipose tissue hormone, adiponectin, seen in obesity, contributes to insulin resistance, atherosclerosis and metabolic syndrome. Patients with metabolic syndrome later on progress into genuine diabetes mellitus.

Obesity

It is the accumulation of excess fat in the body. This in simple terms can occur when the calorie intake is more than the calorie expenditure, the excess calories are deposited as fat. The measurement of obesity can be done by body mass index (BMI) and waist-hip ratio. The BMI is calculated by dividing weight in kg by height in metre square. The normal BMI is between 19-25. A BMI between 25-30 is designated as overweight and a BMI above 30 is obese and persons with a BMI greater than 40 is very obese.

The distribution of fat in the body is different in males (apple type, or android) and in females (gynoid) in females. In males the fat is mainly distributed in the abdominal region and in females fat distribution is in the hips and thighs and hence called as pear shaped. Excessive fat deposition in visceral adipose tissue and abdominal areas are more harmful. The adipocytes of visceral and abdominal region lipolyse faster and the fatty acids enter the portal circulation and reaches the liver. The liver esterifies these fatty acids into TAG which can be incorporated into VLDL along with cholesterol, and secreted into the plasma. The sequential removal of TAG from VLDL by lipoprotein lipase forms first IDL and subsequently, by further removal of more TAG by lipoprotein lipase forms cholesterol-rich LDL. Hence, the upper body adiposity, as seen in males, is more dangerous. This contrasts with the condition obtained in females where lower body adiposity is more prevalent in which the risk posed is less, as these adipocytes are metabolically less active and the fatty acids generated from it by lipolysis goes through the general circulation, and so, can be oxidised in skeletal muscle, so that the liver gets only low supplies of lipids, and hence, the risk is considerably reduced.

The adipose tissue is increasingly being considered as an endocrine organ and it secretes endocrines like leptin and adiponectin. Leptin is secreted by adipocytes, when the fat stored are sufficient and leptin then binds with receptors in the brain to stop feeding. Adiponectin reduces insulin resistance in peripheral tissues.

Due to excess food intake fat cells increase in size and number, and if the adipocytes no longer can take up fat, then the fat may enter other tissues like liver. This can increase insulin resistance. When obese people try to reduce weight by diet restrictions, the number of fat cells are not reduced, but their size only is reduced, meaning, the cells can further attain the original size, if diet restriction is discontinued. A genetic element also seems to be involved in obesity. Further, the maternal nutrition in fetal life plays an important role. The modern amenities have only worsened the state like the sedentary life style and eating high fat and calorie-rich foods, and also the easy availability of tasty food. In short, the problems of overnutrition are on the rise. Regular exercise with dietary restriction, if resorted to meticulously, can bring satisfactory results.

Minor Metabolic Pathways

The minor pathways consists of the following:

(a) Glucuronic acid pathway which produces UDP-glucuronic acid, required for the conjugation of bilirubin, steroids and several drugs. It is also needed for the synthesis of glycosamino glycans. The lower animals can synthesise ascorbic acid with the help of glucuronic acid pathway. The humans cannot synthesise vitamin C, as they lack the enzyme L-gulonate oxidase, making vitamin C and dietary essential.

Pentose is excreted in urine due to the absence of either xylitol dehydrogenase or xylulose reductase, both of which function in the conversion of L-xylulose to D-xylulose. The former gets excreted in urine and as it is having reducing properties, it answers the Benedict's test, and so, should not be mistaken for diabetes mellitus.

(b) Another pathway of significance is sorbitol pathway, which involves the reduction of glucose to sorbitol by aldose reductase. The sorbitol, along with dulcitol are implicated in diabetic cataract development. This also serves to interconvert glucose and galactose. Fructose is present in semen and CSF.

(c) *Fructose metabolism*: Fructose is present in honey and fruits and is acted upon by fructokinase to form fructose 1 phosphate. As it dos not need glucokinase and PFK, fructose is metabolised faster than glucose. As fructose bypasses the glucokinase step requiring insulin, it is tempting to suggest that fructose may be better tolerated by diabetic patients, but, in fact it is not advisable for diabetic patients, because fructose is found to elevate lipogenesis, causing dyslipidemias, creating an atherogenic condition in the patient. The fructose 1 phosphate is converted to dihydroxy acetone phosphate and glyceraldaehyde by aldolase B.

One defect in fructose metabolism is due to the deficiency of fructokinase enzyme known and is known as fructosuria where the benedict's test will be positive. The second Defect is fructose intolerance as a result of the absence of the enzyme aldolase B.

(d) *Galactose metabolism*: Galactose in combination with glucose forms lactose which is present in milk. For the synthesis of galactose, UDP-glucose is epimerised to UDP-galactose. Then, UDP-galactose and UDP-glucose react to form lactose in presence of the enzyme lactose synthase.

Galactose is converted to galactose-1-phosphate by galactokinase. The galactose-1-phosphate then reacts with UDP-glucose to form UDP-galactose and glucose-1-phosphate in presence of the enzyme gactose-1-phosphate uridyl transferase which happens to be the rate limiting enzyme of galactose metabolism. In the next step the UDP-galactose is epimerised to UDP-glucose by the enzyme UDP-galactose epimerase. The enzyme being reversible, galactose can be obtained from glucose and hence, galactose need not necessarily be present in the diet.

Galactosaemia: In this condition the enzyme galactose-1-phosphate uridyl transferase is deficient. The galactose-1-phosphate accumulated will inhibit glycogenolysis, causing hypoglycaemia. Galactosaemia prevails and leads to the formation of dulcitol from galactose. The dulcitol in lens induces osmotic effect, thereby causing cataract. The treatment involves the exclusion of lactose from the diet. The required amounts of galactose for the glycolipids can be obtained by the reversible UDP-galactose epimerase reaction from UDP-glucose.

Another abnormality detected in galactose metabolism is galactokinase deficiency.

chapter 16

METABOLISM OF LIPIDS

DIGESTION OF LIPIDS

The fat content of an average Indian diet is about 20-30 grams a day, whereas the western diet contains much more fat. The fatty acids in the ingested oils may be short, medium, long and very long chain fatty acids or may be saturated, monounsaturated (MUFA) or poly unsaturated fatty acids. The fatty acids may be even chain or odd chain ones. The saliva contains a lingual lipase, but it has no action in the mouth because the optimum pH required for its action is on the acidic side and the pH obtained in the mouth is neutral. The lingual lipase can become active in the stomach where there is an acidic pH prevailing due to the gastric juice containing hydrochloric acid.

The stomach has another lipase, the gastric lipase, which in combination with lingual lipase, both with an optimum pH in the acidic range, acts on the TAGs, with short and medium chain fatty acids, to remove the fatty acids and glycerol. First, the fatty acid, esterified to the hydroxyl group on the first carbon of glycerol is removed to form a fatty acid and 2, 3-diacylglycerol. In the next step, the lipases hydrolyse the ester bond between them at the third carbon of glycerol and the corresponding fatty acid is released leaving 2-mono-acylglycerol. Now the fatty acid on the second carbon of monoacylglycerol is shifted to the third carbon by a mutase enzyme. Next this remaining fatty acid is also cleaved off. The lipid digestion products of lingual and gastric lipases are short and medium chain fatty acids, which are relatively water soluble, and so, can be absorbed directly into blood and not through lacteals as is the case with the long and very long fatty acids requiring the emulsification and micelle formation, needing bile salts. The digestion and absorption of short and medium chain TAGs have to be preceded by emulsification with bile salts. Further, there is no requirement of chylomicron formation in the case of short and medium chain fatty acids. Hence, patients with type-1 hyperlipoproteinaemia, which is due to the absence lipoprotein lipase, involved in the removal of chylomicrons from the plasma, in the postabsorptive state, can be given ghee, butter and coconut oil, all having short and medium chain TAGs, which do not need to form chylomicrons.

The undigested TAGs, having long chain fatty acids, undergo digestion in the small intestine, catalysed by pancreatic and intestinal lipases in presence of colipase. The digestion of lipids with long and very long chain fatty acids should be preceded by emulsification by bile salts (sodium glycocholate and sodium taurocholate), so as to increase the surface area of fat. But for the bile salts, the digestion of fat would have taken many more hours. The pancreatic lipase in presence of colipase hydrolyses the triglycerides into 3 fatty acids and glycerol in a manner similar to the action of gastric lipase, but here optimum pH is alkaline. So, the products of digestion of lipids are fatty acids, mono-acyl glycerol and free cholesterol. The cholesterol esters are hydrolysed by cholesterol esterase and the phospholipase is digested by phospholipase A_2, converting into fatty acids, glycerol, nitrogen base and phosphate.

Next is the formation of micelles, stabilised by bile salts, consisting of fatty acids, mono-acylglycerol and fat soluble vitamins. These micelle are presented to brush border cells of jejunum, which takes up the products of fat digestion into the intestinal mucosal cells, where re-esterification of fatty acids takes place to form TAG and these are packaged into chylomicrons, along with cholesterol ester, phospholipids and fat soluble vitamins and above all apoprotein-B_{48}. As stated earlier, the chylomicrons contain 98 per cent lipid and 2 per cent protein. Out of all the lipoproteins, chylomicrons have least density, as a result of minimum protein content. These lipoprotein particles pass into lacteals and through thoracic duct reach the portal system. The nascent chylomicrons have only one apoprotein, the apo-B_{48}. This further accepts other apoproteins like apo-A, apo-C, apo-E and apo-D. The apo-C_2 is required to activate lipoprotein lipase, as mentioned in the section of lipoprotein metabolism and there is hydrolysis of TAG to fatty acids. Two-three hours after taking means, chylomicrons flood the plasma giving it a turbid look which gets cleared within an hour, thanks to the action of lipoprotein lipase.

The deficiency of pancreatic lipase hampers digestion of fat and hence, unsplit (undigested) fat is excreted in the faeces. On the contrary, the deficiency of bile salts also causes steatorrhoea, where split (partially digested) fat is excreted. An abnormal connection between lymphatic system and ureter can lead to chyluria.

Oxidation of fatty acids

The fatty acids are high energy sources and the oxidation of one mole of palmitic acid produces 106 moles of ATP, whereas the energy output from one molecule of glucose is only 32 ATP moles. The major pathway of oxidation of fatty acid is beta-oxidation. It is so named because of the oxidation taking place at the beta-carbon. The end product of this pathway is acetyl CoA and the coenzymes used are FAD and NAD. The intermediates are attached to coenzyme A. The oxidation process takes place inside the mitochondria.

The fatty acids, before entering the process, have to be activated to acyl CoA derivatives, the enzyme responsible for the activation being acyl CoA synthase. This step takes place in the cytosol. Three different synthase enzymes are seen, one for short chain, second for medium chain and a third for long and very long chain fatty acids. The activation step requires the hydrolysis of 1 ATP to AMP and pyrophosphate, which is equal to 2 ATPs becoming 2 ADPs.

$$\text{Fatty acid} + \text{ATP} + \text{CoA SH} \xrightarrow{\text{Acyl CoA synthase}} \text{Acyl CoA} + \text{AMP} + \text{PPi}$$

The acyl CoA has to be transported from cytosol to mitochondria but it is not permeable through the mitochondrial membrane. Hence, the cell uses a carnitine transport system. The carnitine acyl transferase-1 (CAT-1), located on the outer side of the membrane catalyses the transfer of acyl moiety from acyl CoA to carnitine to form acyl carnitine and CoASH. The acyl carnitine is permeable through the mitochondrial membrane and hence, gets transported to the inside of mitochondria, the translocation across the membrane is effected by carnitine-acylcarnitine translocase. Located on the inside of the mitochondrial membrane is another enzyme, CAT-2, which reforms acyl CoA, which enters the beta-oxidation steps, and carnitine the other product goes back to the outside of mitochondria, to transport more moles of acyl moieties.

The first step in the beta-oxidation of fatty acids is the dehydrogenation of acyl CoA by acyl CoA dehydrogenase to form 2, 3-unsaturated acyl CoA. It is also called as delta-2 transenoyl CoA and alpha-beta unsaturated acyl CoA. The hydrogen atoms released are accepted by FAD to become $FADH_2$ which is oxidised in the respiratory chain to generate 1.5 ATP moles.

In the second reaction, hydration (addition of water) happens on the 2, 3-unsaturated acyl CoA, catalysed by hydratase enzyme to form 3-hydroxy acyl CoA.

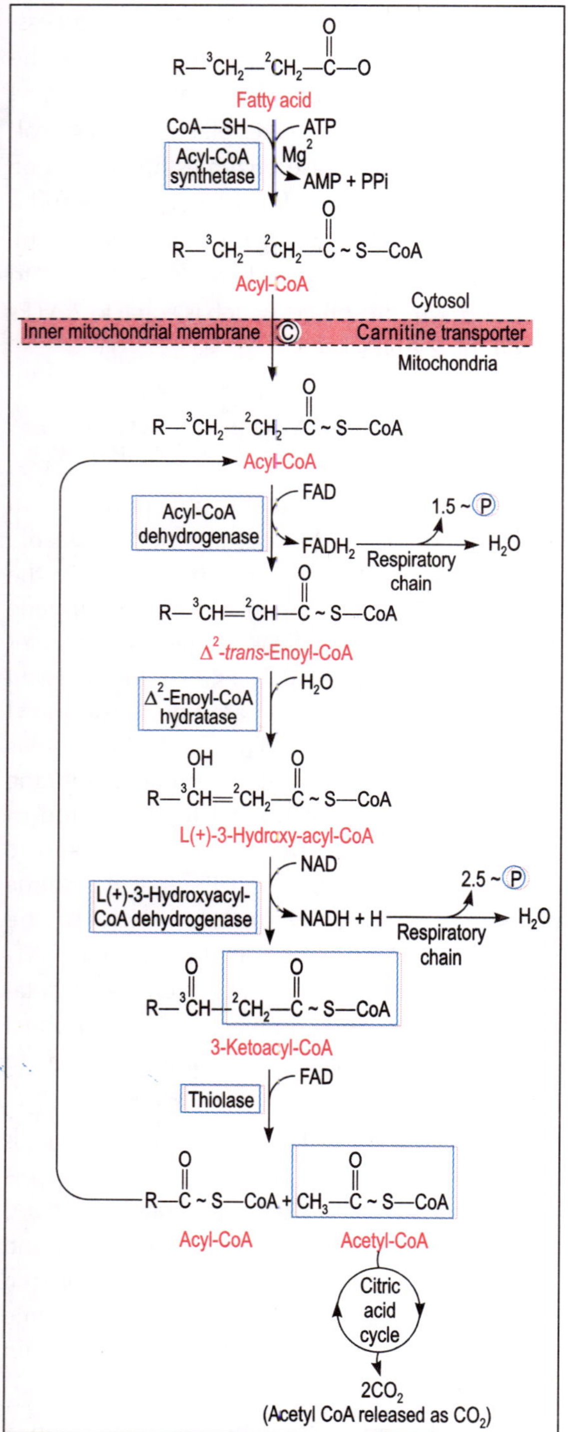

Fig. 16.1: Beta-oxidation of fatty acids

The third step involves a dehydrogenation step catalysed by 3-hydroxy acyl CoA dehydrogenase which generates beta ketoacyl CoA. The reaction requires the presence of NAD to accept the hydrogen and become NADH which gets oxidised in the electron transport chain to produce 2.5 ATP moles. So, till now, 4 ATPs have been produced against the 2 ATPs utilised for the activation of fatty acids.

Next step is the thiolytic cleavage of the beta-ketoacyl CoA by beta-ketothiolase enzyme in the presence of coenzyme A. The products are a two-carbon compound, acetyl CoA and acyl CoA with 2-carbon atoms less. So, if we were to start with palmitic acid (C-16 fatty acid a), the acyl CoA formed at the end of one set of reactions, culminating in the release of one acetyl CoA, shall be having 14 carbons, two carbons being removed as acetyl CoA. The resulting acyl CoA undergoes another round of reactions which further results in the splitting off of another acetyl CoA and an acyl CoA having two carbons less. The process repeats five more times generating a total of 18 acetyl CoA moles. So the total turns required are 7 cycles.

Each cycle produces 4 ATPs and so 7 cycles generate 28 ATPs. The 8 acetyl CoA moles, resulting from the oxidation of palmitic acid, produce 80 ATPs in the TCA cycle (one acetyl CoA produces 10 ATP moles in one turn of the TCA cycle). Therefore, the total number of ATP moles synthesised from one molecule of palmitic acid is 108, but 2 ATPs are used up for the initial activation of fatty acid, and, so, the net synthesis of ATP is 106 in comparison to 32 ATPs produced by one mole of glucose.

The regulation of beta-oxidation is done at the level of CAT-1 which is inhibited by malonyl CoA, which in turn is formed from acetyl CoA catalysed by acetyl CoA carboxylase enzyme involving addition of CO_2, biotin serving as a coenzyme. As it is an energy consuming reaction, ATP is also essential. The acetyl CoA is obtained from glucose through pyruvate. Such a situation prevails in well-fed state, where there is no necessity for fatty acid oxidation to take place. Under such circumstances, lipogenesis is favoured rather than oxidation of fatty acids. The situation is different in fasting and also in diabetes mellitus (no glucose utilisation), which actually leads to more lipolysis from stored deposits of fat releasing fatty acids which undergo beta-oxidation to produce acetyl CoA. Therefore, the hormones glucagon and glucocorticoids (cortisol) stimulates lipolysis, whereas beta-oxidation inhibits lipolysis, and oxidation of fats. In untreated diabetes and starvation, malonyl CoA is decreased, thereby abolishing the inhibitory effect on CAT-1, resulting in elevated beta oxidation which can subsequently lead to ketogenesis and ketoacidosis encountered in diabetes mellitus and starvation.

There are conditions with a decreased rate of beta-oxidation of fatty acids. This can arise due to a deficiency of acyl-CoA synthase. CAT-1, CAT-2 or any other enzymes of the oxidative steps of fatty acids. The unripe fruit of a Jamaican tree known as *akee* contains an inhibitor of acyl CoA dehydrogenase decreasing the catabolism of fat and also causes hypoglycaemia. Further, the deficiency of carnitin can also effect decreased fat catabolism due to defective transport of acyl CoA to mitochondria. This may be encountered in pre-term newborns (decreased synthesis of carnitine by an immature liver) and kidney patients undergoing dialysis (as carnitine may be lost during dialysis). Such subjects may need carnitine administration to take care of hypoglycaemia especially in pre-term infants in whom the feeding is also problematic. The oxidation of fat gives necessary ATPs for the process of gluconeogenesis, especially for the pyruvate carboxylase, and PEP-carboxykinase (GTP is necessary) stages of the pathway. Further, the oxidation of fat also satisfies the requirement of acetyl-CoA to function as an allosteric stimulator of pyruvate carboxylase. Hence, there are chances of hypoglycaemia in premature babies.

What has been described above is the mode of oxidation of even chain saturated fatty acids. The odd chain fatty acids are also oxidised by repeated beta-oxidations, removing acetyl CoA moles, till a three-carbon compound (propionyl CoA) is remaining, which further cannot undergo beta-oxidation. The fate of propionyl CoA is to first form D-methyl malonyl CoA, then L-methyl malonyl CoA, which then gets converted to succinyl CoA. Hence, the last 3 carbons of odd chain fatty acids are glycogenic. A deficiency of vitamin B_{12} can cause methyl malonic aciduria as a result of raised levels of methyl malonic acid in the body especially in the brain causing demyelination leading to neurological disorders.

The unsaturated fatty acids also undergo beta-oxidation till it reaches an existing double bond in a *cis* configuration, which has to be isomerised to the *trans* form, after which the hydratase enzyme adds water to form the corresponding hydroxyl acyl CoA which follows the remaining beta-oxidation steps, as the case with saturated fatty acids. The oxidation of unsaturated fatty acids entails the loss of 1.5 ATP moles for every double bond present. Stated otherwise, for every double bond, the acyl CoA dehydrogenase step involving the reduction of FAD to $FADH_2$ is not necessary and hence, the loss of 1.5 ATPs.

Although beta-oxidation is the major way to oxidise fatty acids, there are other ways of oxidation of fats like alpha-oxidation and omega oxidation and oxidation of very long chain fatty acids in the peroxisomes. The alpha oxidation is more important in the brain, when the diet contains phytonic acids having several methyl groups, all of which have to be removed by alpha-oxidation, prior to the process of beta-oxidation. The phytonic acid is present in milk, chlorophyll, etc. The defect in alpha-oxidation causes Refsum's disease with neurological problems. The alpha-oxidation requires cytochrome P-450. The omega oxidation occurs from the omega carbon which is oxidised to an acidic group to produce dicarboxylic acids that further undergoes beta-oxidation. The very long chain fatty acids cannot be beta oxidised in the mitochondria, and so, are oxidised in the peroxisomes which liberates H_2O_2. A defect in the peroxisomal oxidation of the long chain fatty acids, due to the absence of the necessary enzymes as a result of their important targeting in the peroxisomes, can cause Zell Wager disease (cerebrohepatorenal) which is characterised by accumulation of very long chain fatty acids in brain leading to severe neurological abnormalities which becomes fatal in the early childhood.

Ketogenesis and ketoacidosis

The ketogenesis takes place in the mitochondria, during fasting/starvation and type-1 diabetes mellitus, when the insulin/glucagon ratio is decreased. As there is no intake of food as in starvation, the body uses fat as energy. In type-1 diabetes, there is absolute lack of insulin and such a state also favours lipid utilisation. The fatty acids, produced from lipolysis of triglycerides by hormone sensitive lipase, which is activated by glucagon, produces fatty acids which are oxidised to acetyl CoA by various tissues. Due to the non-availability of oxaloacetate, the acetyl CoA fails to enter TCA cycle, and takes the alternative route of ketogenesis.

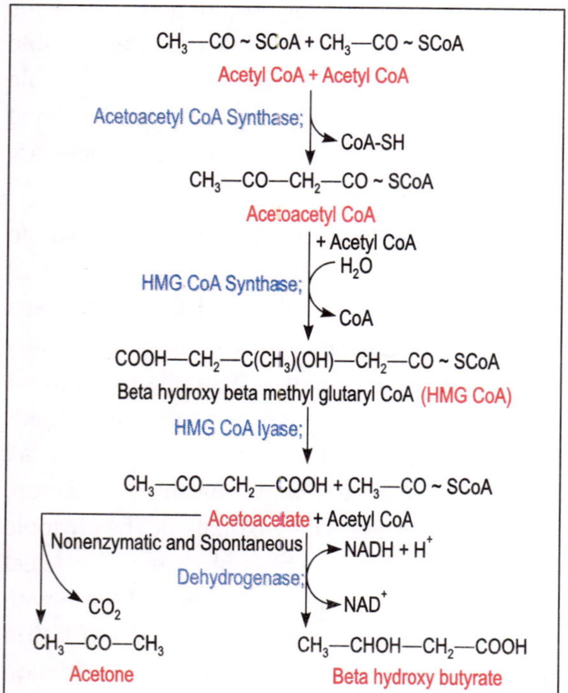

Fig. 16.2: Ketone body formation (Ketogenesis)

Two acetyl CoA molecules condense to form aceto-acetyl CoA in presence of the enzyme aceto-acetyl CoA synthase.

Aceto-acetyl CoA reacts with one more molecule of acetyl CoA and forms hydroxy methyl glutaryl CoA (HMG-CoA), catalysed by the enzyme HMG-CoA synthase.

The HMG-CoA is lysed by HMG-CoA lyase to synthesise acetoacetate and acetyl CoA.

The aceto-acetate is reduced to beta hydroxyl butyrate by dehydrogenase with NADH + H^+ rendering the hydrogen for the reduction.

The aceto-acetate is decarboxylated to acetone by removing CO_2.

Hence, the three ketone bodies are aceto-acetate (primary ketone body), beta-hydroxyl butyrate and acetone (secondary ketone bodies). The ketone bodies are acidic and can cause acidosis, if produced in excess coupled with less utilisation. This is more pronounced in type-1 diabetes mellitus in comparison to starvation. This is more pronounced in type-1 diabetes mellitus in comparison to starvation. The basic phase of blood buffers (sodium bicarbonate) reacts with aceto-acetate to form sodium aceto-acetate and carbonic acid. The carbonic acid is a weak acid as compared to aceto-acetic acid and, hence, the acidity is minimised. The body excretes an acidic urine in its endeavour to control the acidity. By resorting to hyperventilation, the lungs also tries to expel more CO_2. All these mechanisms may be only partially successful as the acetyl CoA production and ketogenesis is in great excess. This leads to ketoacidosis with a decrease in alkali reserve of the body, leading to hyponatraemia and hypokalaemia. The hypokalaemia can cause cardiac arrest. Osmotic diuresis takes place causing dehydration. In diabetes, the administration of insulin, glucose, fluids and electrolytes can better the condition.

After a few days of fasting, the body adjusts itself to the use of ketone bodies as a source of energy except for the brain.

$$\text{Aceto-acetate + Succinyl CoA} \xrightarrow{\text{Succinyl CoA aceto-acetate CoA transferase}} \text{Aceto-acetyl CoA + Succinate}$$

The aceto-acetyl CoA can be split into 2 acetyl CoA by thiolase enzyme. The acetyl CoA can be oxidised in the TCA cycle with minimal amounts of oxaloacetate.

Cholesterol metabolism

The cholesterol a cyclo pentano per hydrophenanthrine ring with 27 carbon atoms. It has 3 benzene rings and one pentane ring with a side chain. It has a hydroxyl group at the third carbon and has a double bond between 5th and 6th carbon. It is the parent sterol from which all the other sterols and steroids are produced. The important compounds produced from cholesterol are comprised of calcitriol, bile acids and steroid hormones. Cholesterol is also a component of cell membranes.

Cholesterol can be synthesised in the body and is not a dietary essential. As it is a reductive biosynthetic pathway, it requires NADPH, and hence, an accompanying HMP shunt pathway. The cholesterol synthesis is very active in liver, adrenal cortex, gonads, etc. These are also the tissue where the pentose phosphate pathway is also active.

The cholesterol synthesis, taking place in the cytosol, starts with the condensation of two-acetyl CoA moles to form acetoacetyl CoA, and is

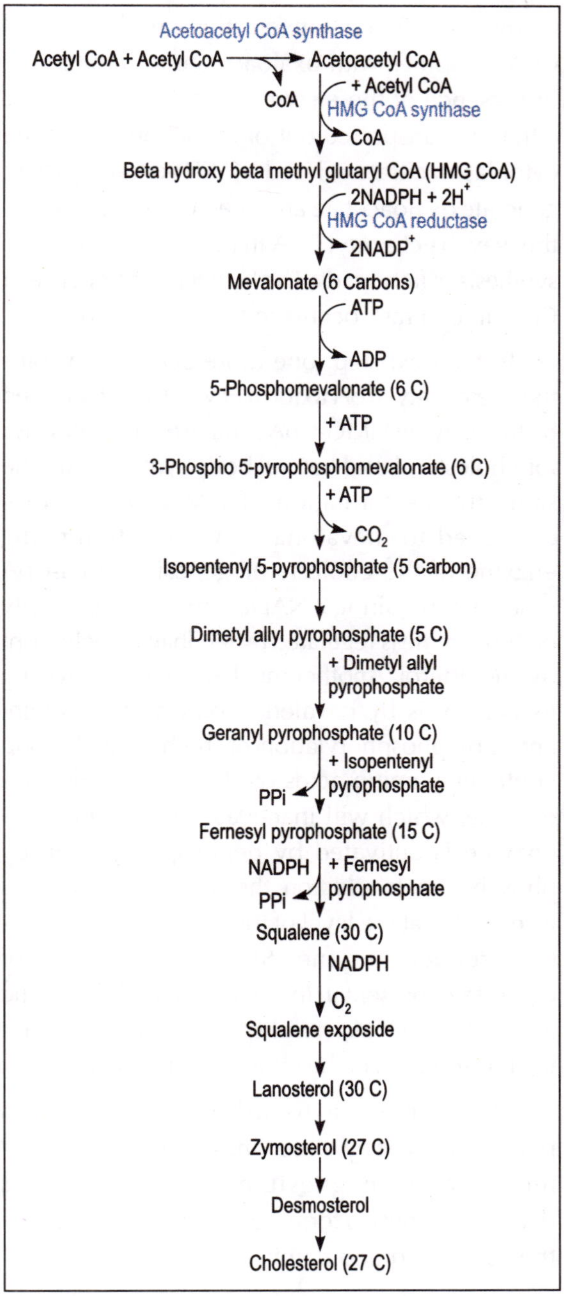

Fig. 16.3: Cholesterol biosynthesis

catalysed by the reversal of the thiolase enzyme. The acetyl CoA is produced in the mitochondria from pyruvate by PDH complex, but the cholesterol synthesis occurs in the cytosol. Hence, the acetyl CoA has to be transported to the cytoplasm, but it is not permeable through the membrane. To circumvent this problem acetyl CoA combines with oxaloacetate to form citrate in presence of the enzyme citrate synthase. The citrate is transported out of mitochondria. In the cytoplasm, citrate is lysed by ATP-citrate lyase to generate oxaloacetate and acetyl CoA. This is also the way to get acetyl CoA in the cytoplasm for the synthesis of fatty acids. The ketogenesis from acetyl CoA, in contrast, occurs in the mitochondria.

In the next step, one more acetyl CoA joins the acetoacetyl CoA to form HMG-CoA, catalysed by the enzyme HMG-CoA synthase. The pathways for cholesterol and ketone body synthesis are the same till the formation of HMG-CoA. This is converted to mevalonate by the rate-limiting enzyme of the cholesterologenesis, HMG-CoA reductase, requiring 2 NADPH moles. The activity of the enzyme is regulated by feedback mechanism by cholesterol. Another mechanism of controlling its activity is by covalent modification brought about by phosphorylation of the hydroxyl group containing amino acids of HMG-CoA reductase enzyme, which will inactivate the enzyme. The enzyme is activated by dephosphorylation by phosphatase. In addition, the cholesterol synthesis is regulated at the level of the expression of HMG-CoA reductase gene. Sufficient amounts of cholesterol present in the cells can bind to specific areas on the DNA, sterol specific sites, to inhibit the transcription of the HMG-CoA reductase gene.

Mevalonate, a 6-carbon compound, is phosphorylated by three kinases involving 3 ATP moles and magnesium, and undergoes decarboxylation to form isopentenyl diphosphate, the active isoprenoid unit.

Squalene is formed from 6 isoprenoid units. One isopentinyl diphosphate is isomerised to dimethylallyl diphosphate which condenses with another mole of isopentenyl diphosphate to form a ten-carbon geranyl diphosphate. The geranyl diphosphate combines with another molecule of isopentenyl diphosphate to synthesise the farnesyl diphosphate, a 15-carbon compound. Two such farnesyl diphosphate condense to generate squalene, a 30-carbon compound, the formation of which involves the elimination of 2 inorganic pyrophosphates and reduction with NADPH.

Subsequently, the squalene becomes lanosterol by shifting the methyl groups from C_{14} to C_{13}, and from C_8 to C_{14}. This conversion of squalene to lanosterol also requires 2 NADPH molecules and one FAD in addition to oxygen. The enzymes needed include squalene epoxidase and cyclase.

The lanosterol undergoes several modifications to become cholesterol. The methyl groups on C_{14} and C_4 of lanosterol are removed as CO_2 to become zymosterol. This also requires 2 more molecules of NADPH, NAD and oxygen.

The double bond between C_8 and C_9 is now transferred to between C_5 and C_6, thereby forming desmosterol. This molecular change also needs NADPH and oxygen.

The double bond in the side chain of desmosterol is reduced with NADPH to form the final product, cholesterol. A total of 8 NADPH molecules are necessary for the synthesis of one molecule of cholesterol.

The cholesterol as such or after conversion to bile acids is excreted by liver through the bile, which is the only way of excreting cholesterol from the body.

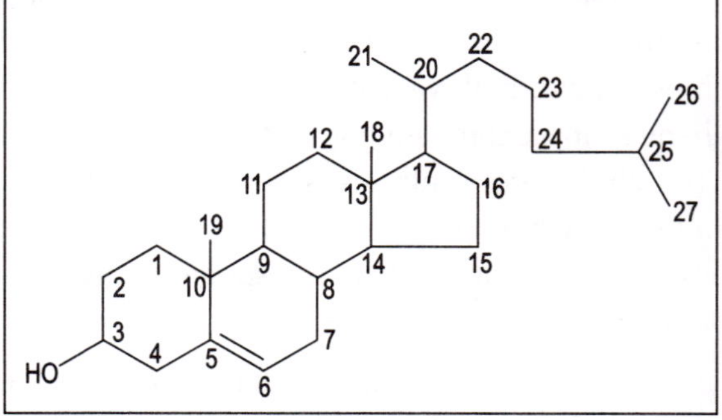

Fig. 16.4: Cholesterol

The serum cholesterol level is correlated with the incidence of atherosclerosis and coronary heart disease. A serum total cholesterol level of 150 to 200 mg/dl is considered to be normal and values around 220mg/dl is thought to be a borderline case. Values above 250mg/dl definitely is linked to atherosclerosis and CAD.

Taking more PUFA, MUFA and omega-3 fatty acids more common in a vegetarian diet helps in decreasing the not so good LDL and total cholesterol but increasing the healthy HDL-cholesterol because LDL is atherogenic and HDL is antiatherogenic. High blood pressure, male gender, smoking, coffee intake, emotional stress, lack of exercise and abdominal obesity can also cause heart ailments.

When the dietary modifications fail to give the desired levels of serum lipids, hypolipidemic drugs like stains can be administered. These drugs inhibit the HMG-CoA reductase enzyme, the rate limiting enzyme of cholesterologenesis. Ezetimibe inhibits cholesterol absorption from the intestine. Clofibrates reduce the serum TAG levels. Bile acid binding resins complex with bile salts and prevent the entero-hepatic cycle. Now more cholesterol is converted to bile salts and this leads to decrease in cholesterol levels.

Metabolism of Lipoproteins

The lipoproteins are complexes of lipids and proteins. Lipids, like triacylglycerol (TAG), cholesterol and phospholipids are not soluble in the aqueous plasma and so cannot be transported on its own. The lipids have to be complexed with proteins so as to be transported in the plasma. The lipoproteins can be classified into the following groups: chylomicrons, VLDL (very low density lipoproteins), IDL (intermediate density lipoproteins), LDL (low density lipoproteins) and HDL (high density lipoproteins). Although not a true lipoprotein, fatty acid-albumin complex is another related substance.

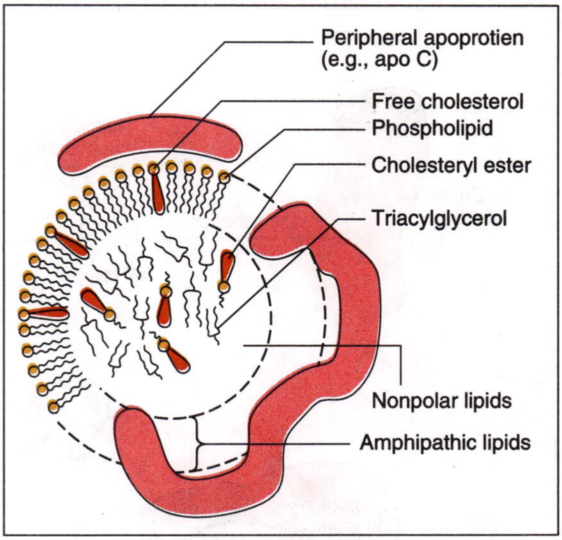

Fig. 16.5: Lipoprotein

Chylomicrones

Before the chylomicron formation, in the intestinal mucosal cell, the dietary lipids are digested by the combined action of lingual, gastric (stomach), pancreatic and intestinal lipases (small intestine). Before the process of digestion in the intestine, the emulsification of fat has to take place with the help of bile salts. The products of digestion of lipids, fatty acids, cholesterol, monoacyl glycerol and fat soluble vitamins are formed into micelle which is absorbed into the intestinal mucosal cells.

The chylomicrons contain dietary lipids and represent exogenous lipids and are formed in the intestinal mucosal cells. The protein content of chylomicrons is 2 per cent and the lipid content is 98 per cent. The lipids present in this lipoprotein are a maximum of TAG and small amounts of cholesterol and phospholipids. The chylomicrons also have little amounts of cholesterol and phospholipids. The chylomicrons also have little amounts of fat soluble vitamins. The protein part of lipoproteins is known as apoprotein and the apoprotein of chylomicron is apoprotein B_{48} which is 48 per cent of apoprotein B_{100}, seen in VLDL, IDL and LDL. The chylomicrons enter the

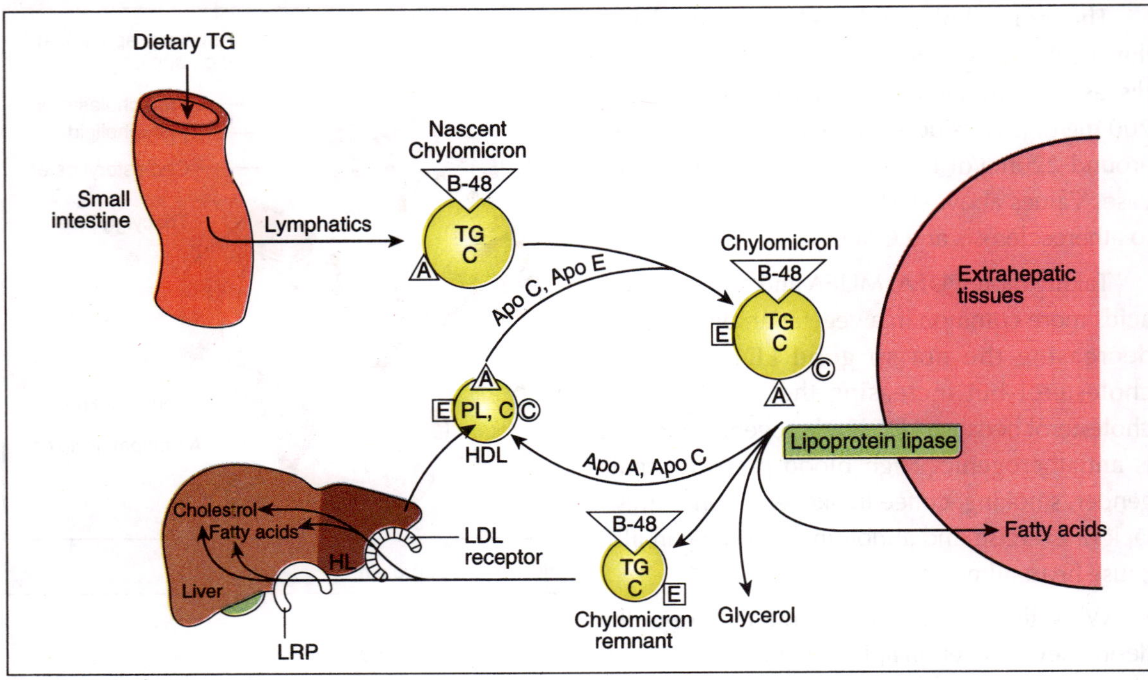

Fig. 16.6: Metabolism of chylomicrons

lymphatic system through lacteals and then go to the thoracic duct and then to the blood, thereby, making the plasma turbid or milky due to the presence of chylomicrons which are removed from plasma within 1-2 hours by the action of lipoprotein lipase which is located in the walls of the capillaries of muscles and adipose tissue. The nascent chylomicrons formed have only apo-B_{48} apoprotein. When it reaches the blood it receives apo-C, apo-E and apo-D from the reservoir or repository of apoproteins, the HDL. With this the nascent chylomicrons become mature ones. The lipoprotein lipase is activated by apo-C_2 and this initiates the hydrolysis of TAG present in the chylomicrons to fatty acids and glycerol, the fatty acids going to the tissues. The step by step removal of TAG from the chylomicrons make it relatively less in TAG, but more in protein making the chylomicron remnants. These remnants may be taken up by tissues and can be degraded by the enzymes present in the cells or may be further hydrolysed by lipoprotein lipase.

Very low density lipoproteins (VLDL)

The VLDL, rich in TAG, are formed in the liver and hence, carry the endogenous TAGs along with cholesterol and phospholipids. The triglycerides and cholesterol esters, being hydrophobic, are at the central core, and the phospholipids and apoproteins are at the periphery of the molecule. The VLDL has about 10 per cent protein and 90 per cent lipid. The nascent VLDL has only apo-B_{100}. When VLDL is secreted into the blood, it receives from HDL, apo-C, apo-E and apo-D to become mature VLDL. The metabolism of VLDL bears similarities to the metabolism of chylomicrons. The lipoproteins lipase, activated by apo-C_2, present in the capillary walls of peripheral tissues, hydrolyses the triglycerides to fatty acids and glycerol in a manner analogous to chylomicrons. After the initiation of lipoprotein lipase action, the apo-C and apo-D goes back to HDL. By the sequential removal of TAG by lipoprotein lipase, the content of TAG gradually

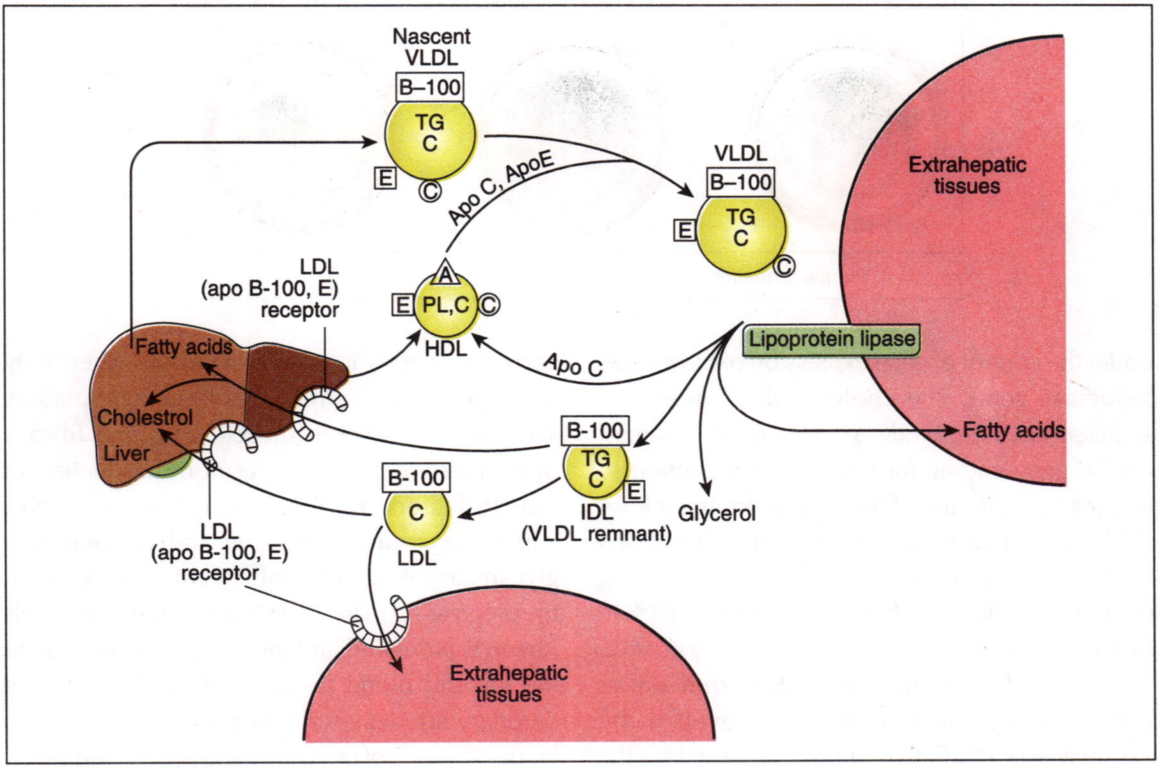

Fig. 16.7: Metabolism of VLDL

declines with a relative rise in cholesterol and protein levels thereby generating intermediate density lipoproteins (IDL). The IDL have protein content of 16 per cent with the lipid content at 84 per cent. A part of the IDL can be taken up by peripheral tissues by apo-E specific receptors and gets degraded in the cells. At this point of time, the apo-E also returns to the HDL. The IDL is further acted upon by lipoprotein lipase, removing the remaining TAG so as to become the cholesterol-rich low density lipoproteins (LDL) with only apo-B_{100}.

Low Density Lipoproteins

The LDL formed from VLDL through IDL by the action of lipoprotein lipase is taken by LDL-receptors located in clathrine coated pits situated on the cell membrane of peripheral tissues. The LDL particles bind to the apo-B_{100} specific LDL-receptors and the complex is internalised by endocytosis. The LDL-receptors return to the cell surface to bring more LDL. The endosomes containing LDL combines with lysosomes. The apo-B_{100} is degraded by lysosomal proteases to amino acids. The cholesterol is used for the formation of steroid hormones, bile acids and calcitriol or may become a part of cell membrane or may get esterified and stored. The cholesterol, absorbed like this, can inhibit the synthesis of cholesterol by competitively inhibiting the rate limiting enzyme of cholesterol biosynthesis, hydroxy methyl glutary CoA reductase (HMG CoA reductase). The cholesterol taken up can also down-regulate the number of LDL-receptors, decreasing the further uptake of cholesterol. Therefore, the uptake of LDL through LDL-receptors is a regulated process and is under feedback control. The cholesterol synthesis is also

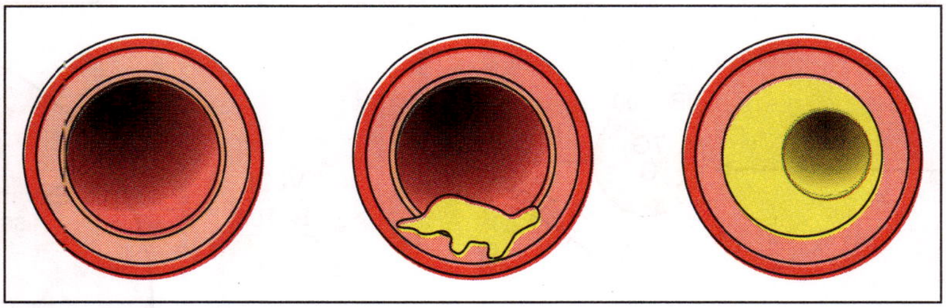

Fig. 16.8: Plaque formation

under the control of gene expression of HMG CoA reductase gene. The cholesterol, if present in sufficient amounts, binds to sterol specific elements on the gene coding for HMG CoA reductase so that the transcription of the corresponding mRNA is blocked, and, hence, its translation to HMG-CoA reductase enzyme, thereby regulating cholesterol synthesis. Hence, the cholesterol taken up by peripheral tissues exerts a three-way control, first on its further uptake and second on its synthesis and third at the gene level in the expression of HMG-CoA reductase gene into the enzyme. The rate-limiting enzyme of cholesterol synthesis, HMG-CoA reductase can be regulated by covalent modification by phosphorylating it through a mechanism involving cyclic AMP. The phosphorylated enzyme in this case is inactive, whereas the active form is the phosphorylated variety.

Cholesterol uptake by macrophages: An unregulated process

Although the uptake of LDL by LDL receptors is a finely regulated process with the up or down regulation of LDL receptors, the uptake of LDL, oxidised LDL and glycosylated LDL (in diabetes mellitus) by macrophages is a nonspecific and unregulated process, and so, can result in excess uptake of LDL by the macrophages to become cholesterol laden foam cells. These cells accumulate more cholesterol and goes to the subintimal layer of the arterial walls of aorta, coronary arteries and cerebral vessels, where they get deposited to form an atherosclerotic plaque. This leads to smooth muscle cell, and fibrous proliferation ensues as a result of production of growth factors by platelets and macrophages, which accumulates collagen and glycosamino-glycans resulting in an inflammatory state. With the progression in the plaque formation, the vessels lumen is narrowed and thrombosis occurs due to the forceful blood flow. The thrombosis in the coronary arteries results in ischaemia in cardiac tissue. The lack of oxygen supply leads to anerobic glycolysis, generating minimal ATPs, insufficient functioning of cardiac tissue and death (necrosis) of the tissue resulting in myocardial infarction which can be diagnosed by ECG changes and assaying serum troponins and CK MB concentrations. A block in the cerebral vessels can cause stroke.

The LDL cholesterol is important in the transport of cholesterol from liver to peripheral tissues. The atherosclerosis starts only with increased level of LDL more than 130mg/ml and total cholesterol more than 220mg/dl. The LDL cholesterol is considered to be "bad cholesterol" if it exceeds normal levels.

High Density Lipoproteins

The HDL is formed in the enterocytes and released into the blood. Initially it contains free cholesterol, phospholipid, lecithine and apoproteins A, C, D and E. The HDL collects cholesterol from the

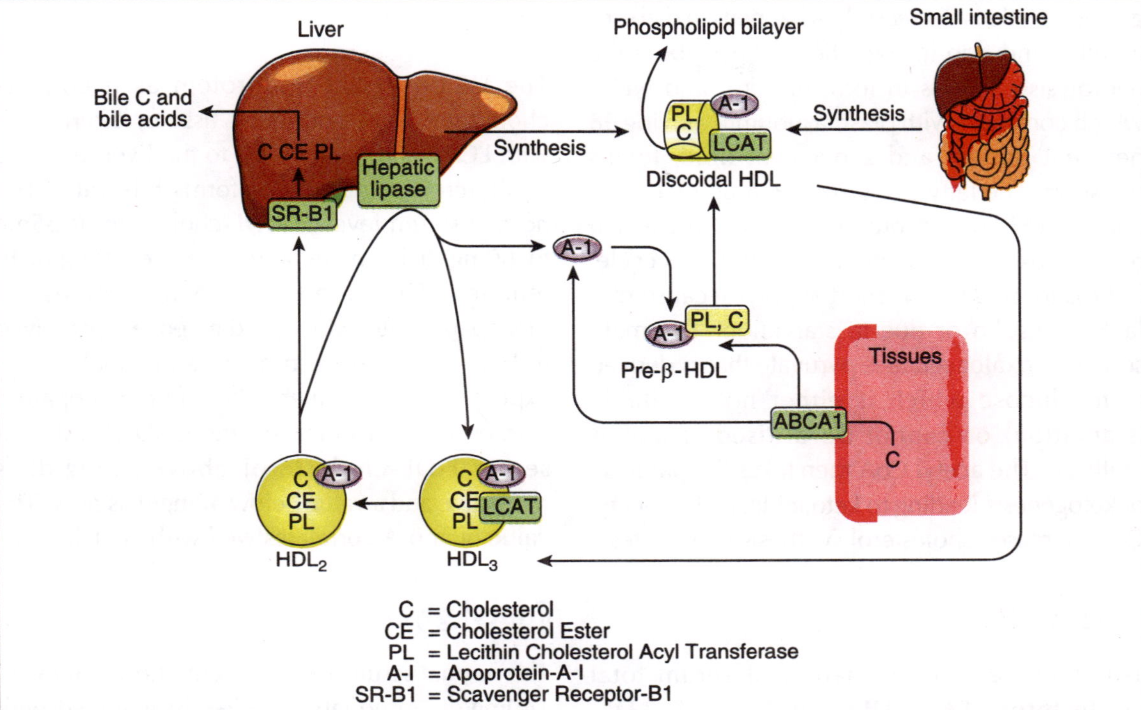

Fig. 16.9: Metabolism of HDL

peripheral tissues (scavenging function) with the help of ATP cassette binding protein, the absence of which causes Tangier disease, in which there is decreased level of alpha lipoproteins in the plasma and such patients develop early atherosclerosis. The HDL is associated with an enzyme lecithin cholesterol acyl transferase (LCAT) which acts on free cholesterol and the phospholipid lecithin and transfers the poly unsaturated fatty acid (PUFA) positioned on the second carbon of the glycerol of lecithin to the hydroxyl group of free cholesterol to form cholesterol ester. This is an important step towards cholesterol transport to liver and forms HDL 3 with apo-D. The HDL 3 accepts more cholesterol, esterifies it and forms HDL 2 which is antiatherogenic. Hence, the PUFA helps in the transport of cholesterol in association with HDL. The HDL 2 consists of HDL 2A and HDL 2B. The antiatherogenic apo-J is seen along with HDL 2.

The cholesterol ester transfer protein, associated with HDL, transfers cholesterol ester from HDL to other lower density lipoproteins, like VLDL, IDL and LDL in exchange of triglycerides.

Dyslipidemias

Both hyper and hypolipidemias are included in this. In Tangier disease alpha lipoproteins are reduced. Fredrickson classified familial hyperlipidemias into five types, out of which Type IIB is most dangerous as both cholesterol ANA TAG are increased. Secondary hyperlipidemias are seen diabetes, obstructive jaundice and nephrotic syndrome.

Fatty acid-albumin complex

Although, not a lipoprotein in the true sense, the nonesterified fatty acids (NEFA) combines with plasma albumin to form a complex for transport. The NEFA are formed by the action of hormone sensitive lipase on adipose tissue TAG. The hormone sensitive lipase, stimulated when insulin glucagon ratio is decreased, as happens in diabetes mellitus and also in starvation, brings about the hydrolysis of TAGs to form glycerol and NEFA which combines with plasma albumin, having 98 per cent protein and 2 per cent lipid, for its transport. The fatty acids are taken up by tissues and are oxidised to produce acetyl CoA which requires oxaloacetate for entering the TCA cycle, but oxaloacetate is in short supply because of a lack of insulin or due to starvation. The main source of oxaloacetate is pyruvate that is derived from glucose which is either not available (starvation) or cannot be utilised (diabetes mellitus). The acetyl CoA then takes the pathway of ketogenesis leading to ketoacidosis. The acetyl CoA increases cholesterol synthesis in diabetes.

Lipid profile

This includes the estimation of serum total cholesterol, TAG, HDL-cholesterol, LDL-cholesterol, and VLDL-cholesterol.

Total cholesterol

The total cholesterol consists of HDL, LDL and VLDL cholesterol. The normal serum cholesterol value is 150-200mg/dl. A value of cholesterol more than 200mg/dl but less than 220mg/dl is said to be borderline and cholesterol concentrations between 220-250mg/dl carries moderate risk. If the cholesterol level is more than 250mg/dl, there is real risk of developing CAD.

LDL-cholesterol

The normal serum LDL-cholesterol level is 70-130 mg/dl. A value higher than 170 is dangerous and is positively correlated with atherosclerosis and CAD. The LDL in excess is bad because it transports cholesterol from the liver to the peripheral tissues and can cause atherosclerosis and CAD. The concentration of Apo-B_{100} correlates well with LDL concentrations.

HDL-cholesterol

The HDL is good lipoprotein as it collects cholesterol from peripheral tissues, esterified it with LCAT and transports it to the liver, excretes cholesterol in the bile or forms bile salts. The normal serum level of HDL-cholesterol is 35mg to 60mg/dl in males and 40mg to 70mg/dl in females. This trend is seen in females till menopause, after which no difference is perceived in HDL levels between males and females. This explains the less vulnerability of premenopausal women to atherosclerosis and CAD. A value of serum HDL-cholesterol above 55mg/dl is protective and a value below 40mg/dl is risky. The value of apo-A correlates well with HDL levels.

Triacyl glycerol

The normal serum TAG concentration is 50mg to 140mg/dl. Although not directly involved with atherosclerosis, raised levels of TAG can cause insulin resistance, adversely affecting health.

Lipoprotein 'a'

This has atherogenic properties and it has resemblance to tissue plasminogen, which results in the inhibition of fibrinolysis. This in turn can cause thrombosis and atherosclerosis. So raised levels of lipoprotein 'a', remaining associated with LDL is implicated in heart ailments. The Indians, in general, having higher levels of lipoprotein 'a'.

Non-HDL cholesterol

This term is used to describe the total of LDL, VLDL, IDL and lipoprotein 'a', the value of which should not exceed 130mg/dl, a value of 130-160mg/dl is borderline and above 160mg/dl carries high risk.

Total cholesterol/LDL-cholesterol ratio should not exceed 3.5. Another important factor is the ratio of LDL-cholesterol to HDL-cholesterol which should not exceed 2.5.

Risk factors for Cardiovascular disorders

Diabetes mellitus increases the risk of cardiovascular disorders due to factors like insulin resistance, hypercholesterolaemia, elevated glycosylated and/or oxidised LDL and obesity.

An HDL level below 35mg/dl is linked to cardiovascular complications, whereas an LDL value above 170mg/dl is also dangerous.

As mentioned elsewhere, a ratio of total cholesterol to HDL-cholesterol should not be higher than 3.5 and the ratio of LDL to HDL should not exceed 2.5.

The lipoprotein 'a' should not be more than 30 mg/dl.

Yet another risk factor is raised levels of homocysteine which can replace glycine from collagen, thereby leading to the exposure of collagen in the blood vessels resulting in thrombus formation. This culminates in the development of atherosclerosis and CVD. Therefore, the level of homocysteine should remain within 15 micromole/L.

High blood pressure causes CVD and cerebral stroke.

Obesity which is alarmingly on the rise is linked to CVD. Hence, weight reduction by regular exercise coupled with healthy eating should be resorted to.

Cigarette smoke contains nicotine which is vasoconstrictive and can be risky.

High-sensitive C-reactive protein is a predictor of future cardiovascular accidents.

Prostaglandins

The prostaglandins, first isolated from prostate glands and seminal fluid, are obtained from the membrane-bound arachidonic acid. The phospholipase A_2 acts on membrane phospho-

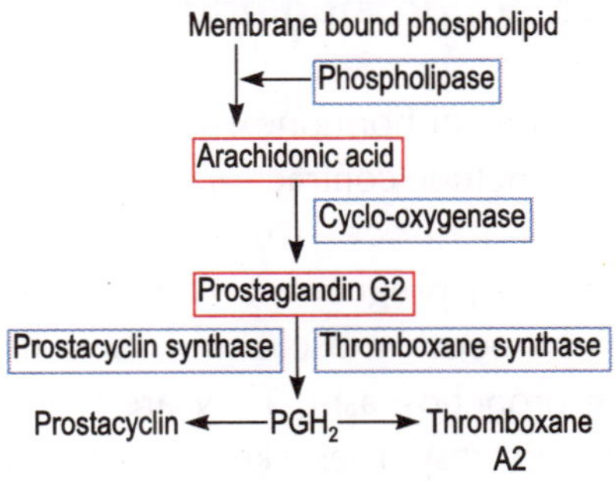

Fig. 16.10: Synthesis and action prostaglandins

lipids to form arachidonic acid. The arachidonic acid can be acted upon by cyclo-oxygenase or by lipo-oxygenase. The cyclo-oxygenase acts on arachidonic acid to form prostaglandin G_2, which converted by peroxidise to prostaglandin H_2. The PG H_2 is converted by thromboxane synthase to thomboxane A_2. The prostacyclin (PG_{12}) is formed by prostracyclin synthase from prostaglandin H_2. The thromboxane A_2 causes platelet aggregation and vasoconstriction, whereas the prostacyclins are responsible for preventing platelet aggregation and cause vasodilation.

The leukotrienes are produced by the lipoxygenase pathway from arachidonic acid

The cyclogenase is inhibited by nonsteroidal antiinflammatory drugs. Aspirin acts as a suicidal inhibitor of cyclo-oxygenase. The enzyme action on aspirin makes it more potent which irreversibly inhibits the enzyme, hence the name suicidal inhibition. The aspirin is beneficial to heart patients because aspirin, irreversibly inhibits the platelet cyclo-oxygenase which cannot be synthesised again resulting in decreased formation of thromboxane A_2 which results in less aggregation of platelets, thereby helping the heart patients. The endothelial cyclo-oxygenase, on the other hand, can be regenerated, and so, the PG_{12} can be synthesised. There are two forms of cyclo-oxyenase, of which the reducible form is responsible for the inflammatory response. The half-life of PGs is only 30 seconds.

Functions of prostaglandins

They are as local hormones. Their functions are varied and include contraction of uterine smooth muscles and hence, finds use in medical termination of pregnancy. The PGs relax the bronchial smooth muscle and so is useful in relieving bronchospasm. They are mediators of inflammatory responses like pain and swelling. Thromboxane A_2 causes platelet aggregation and vasoconstrictions. The prostacyclins have the opposite effects, that is, preventing platelet aggregation and causing vasodilatation. PGE-2 decreases blood pressure. It also inhibits gastric secretion. The prostaglandins act by decreasing cyclic AMP and prevents lipolysis.

Leukotrienes, another group of prostanoids, are chemotaxic agent. They are also responsible for allergic reactions as the slow reacting substance of anaphylaxis (SRS) contains leukotriene C_4, LTD_4 and LTE_4. The SRS is involved in asthma and other hypersensitivity reactions. Another group of related compounds are lipoxins which lessens immune response.

chapter 17

METABOLISM OF PROTEINS

The protein metabolism consists of digestion and absorption, general reactions in the metabolism of amino acids like transamination, deamination, decarboxylation, detoxification of ammonia by forming urea, special products formed from amino acids and inborn errors of amino acid metabolism.

Digestion of Proteins

The digestion of proteins takes place in the stomach and the small intestine. The gastric juice contains the inactive pepsinogen which is activated by HCl. For the activation, a peptide of about 45 amino acids is removed from the amino terminus of pepsinogen when it becomes the active enzyme pepsin which specifically acts on peptide linkages having phenylalanine and tyrosine. It does not act on structural proteins. The optimum pH of pepsin is about 2. Rennin is an enzyme present in infants for the curding of milk especially acting on casein. Pepsin converts proteins into peptones and proteoses and is an endopeptidase.

The pancreatic juice contains trypsinogen, chymotrypsinogen and procarboxypeptidase. The inactive trypsinogen has to be activated to trypsin by the enzyme enterokinase by removing 6 amino acids from the N-terminus. The trypsin has an optimum pH in the range of 8-9. It acts on the products of digestion of pepsin and degrades the peptones and proteoses to peptides and ultimately to amino acids. It is also an endopeptidase. The trypsin acts on the peptide bonds whose carboxy groups are contributed by basic amino acids.

The pancreatic juice also contains the inactive chymotrypsinogen which is to be activated to chymotrypsin, the activation carried out by trypsin. It is also an endopeptidase having an optimum pH of 8-9 and is similar in its action to trypsin.

The exopeptidases are amino peptidase and carboxy peptidase, the former acts from the aminoterminus and the latter acts from the carboxyterminus. These peptidases are also secreted as inactive enzymes, like procarboxy and proamino peptidases which are activated respectively to carboxy and amino peptidases.

The activation of inactive proteolytic enzymes to active forms is due to post-translational modifications. The secretion of protein digesting enzymes as inactive proenzymes will avoid autodigestion of intestinal mucosa.

Absorption of amino acids

The absorption of amino acids requires energy. There are different carriers for different classes of amino acids like one carrier for neutral amino acids, second for basic ones, third for acidic ones and a fourth for glycine and proline.

The end products of digestion of proteins are amino acids which being soluble can be absorbed into the blood and then transported to the liver. The naturally occurring amino acids are of L-isomer type. The amino acids also reach all the tissues and are used for various functions.

The food allergy can be caused by the absorption of partially digested peptides which can elicit immune response and this may explain the allergy due to milk and fish ingestion in certain individuals.

Certain inborn errors of amino acid metabolism may have roots in the defective transport of amino acids. Examples are cystinuria and the disease regarding the metabolism of tryptophan and cystic fibrosis.

In the newborn infants, the immunoglobulins of the mother's milk can be directly absorbed without digestion and without producing any allergic reaction.

Amino acid pool

The body proteins undergo turnover and the amino acids resulting from it mix with the dietary ones to form an amino acid pool from which amino acids are withdrawn for the synthesis proteins as different varied turnover times. The catabolism of proteins is accelerated by glucocorticoids. The amino acids remaining after protein synthesis is transaminated or deaminated and may be used for the synthesis of various special products. Amino acids are not stored in the body. In the plasma, the amino acids in highest concentration are glutamic acid and glutamine while the remaining amino acids are low.

In a fasting state, the degradation of muscle proteins occurs due to the influence of corticoid hormone and the amino acids which are taken up by the liver and deaminates the amino acids to form keto acids like pyruvate and ammonia. The keto acids are converted to glucose and the ammonia detoxified to urea. There is an alanine-glucose cycle occurring between muscle and liver. The amino acids can be taken up by different tissues and can be employed in various synthetic processes.

General metabolic pathways of amino acids

1. The primary function of amino acids is to synthesise proteins.
2. Another function of amino acids is to take part in the synthesis of special products like creatine, thyroxine, haem, carnitine, nucleobases, GABA, etc.
3. Several of the amino acids undergo transamination and deamination and the resultant ammonia, being toxic, is converted to urea in the liver and excreted in the urine.
4. Another function is the synthesis of nonessential amino acids from the ketoacids which are produced by the transdeamination process of amino acids.
5. The gluconeogenesis can take place from the carbon skeletons of various glucogenic amino acids.
6. The decarboxylation of amino acids can form the corresponding amines. For example histidine can be decarboxylated to histamine, tyrosine is converted to tyramine, tryptophan to tryptamine and 5-hydroxy tryptamine (serotonin, causing relaxation), and glutamate to gamma amino butyrate, a neurotransmitter.
7. Amino acids, although primarily meant for body protein synthesis, can also be used for generating energy by getting converted to TCA cycle intermediates.

Transamination

The transamination is the process in which there is an exchange of the alpha amino group between a pair of amino acids and keto acids. The enzymes catalysing these reactions are transaminases or aminotransferase and require pyridoxal phosphate as a coenzyme. Except for lysine, threonine and proline, all the amino acids can undergo transamination. The following reactions are two important transamination reactions. In the first reaction alanine and alpha-ketoglutarate react in presence of alanine aminotransferase and pyridoxal phosphate to synthesise glutamate and pyruvate. In these reactions the pyridoxal phosphate combines with the amino group of the amino acid to generate pyridoxamine phosphate.

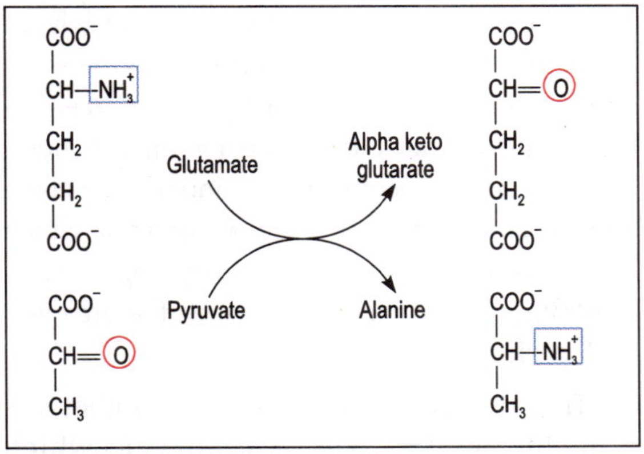

Fig. 17.1: Transamination reaction

The pyridoxamine phosphate, then, transfers its amino group to a ketoacid to form the corresponding amino acid. This mechanism of enzyme action is known as ping-pong mechanism.

Alanine aminotransferase (ALT)/
Pyridoxal phosphate

Alanine + alpha-keto glutarate ⟷ Pyruvate + glutamate

In this reaction, as it is reversible, it is possible to form either pyruvate or glutamate depending on which amino acid (alanine or glutamate) is in short supply. If there is scarcity of glutamate, the forward reaction will be favoured and if alanine is less, the backward reaction will be more active.

In the second reaction, aspartate and alpha-ketoglutarate react in presence of aspartate aminotransferase (AST) and pyridoxal phosphate to form glutamate and oxaloacetate.

Aspartate aminotransferase (AST)/
Pyridoxal phosphate

Aspartate + alpha-keto glutarate ⟷ Oxaloacetate + glumate

Here again, the reaction being reversible, either aspartate or glutamate can be synthesised, depending on the scarcity of aspartate or glutamate, rspectively, as the case may be.

The ALT is also called as glutamate pyruvate transaminase (GPT) and the AST is also designated as glutamate oxaloacetate transaminase (GOT). Hence, it is possible to synthesise the nonessential amino acids.

Both the ALT and AST serve as important tools for the diagnosis of hepatocellular jaundice and cirrhosis of liver.

Oxidative deamination

Oxidative deamination taking place after transamination is also termed as transdeamination. The amino groups of most of the amino acids are finally concentrated in glutamic acid by the process of transamination. The glutamic acid then undergoes oxidative deamination to alpha-ketoglutarate and ammonia.

Glutamate dehydrogenase

Glutamate + NAD + H_2O ⟷ alpha-keto-gluterate + ammonia + NADH

There are other amino acid oxidases which use either FAD or FMN. The non-oxidative deamination removes ammonia from serine while forming pyruvate.

Disposal of ammonia

The ammonia being a toxic substance, especially to the brain, has to be detoxified immediately. In the brain, glutamic acid combines with ammonia in presence of ATP to form glutamine, the reaction being catalysed by the enzyme glutamine synthase. The glutamine after reaching liver is converted back to glutamate by another enzyme glutaminase. The glutamine in the kidneys is of special significance as it is acted upon by glutaminase producing ammonia which is involved in the acid base balance because ammonium ion replaces sodium and potassium from salts excreted in the urine in acidosis. Glutamate, glutamine and alanine serve to transport ammonia from other tissues to liver which is the final site of detoxification of ammonia.

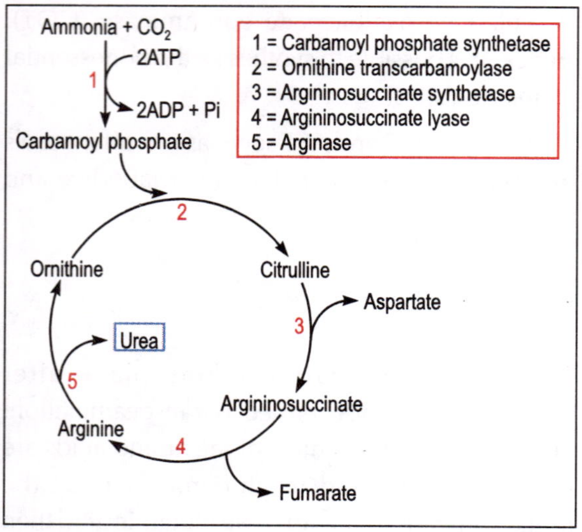

Fig. 17.2: Urea cycle

Urea formation

The main pathway to detoxify ammonia is to form urea. It was elucidated by Hans Krebs and Henseleit and hence, is known after them. The urea cycle is also the first metabolic pathway to be discovered.

In the first step of urea cycle one molecule each of ammonia and CO_2 condense in the presence of carbamoyl phosphate synthase-1 (CPS-1) and 2ATP molecules to form carbamoyl phosphate. This reaction occurs in the mitochondria and CPS-1 is the key enzyme of urea cycle and is hence irreversible. The two ATP molecules are hydrolysed to 2ADPs and 2 phosphates. The N-acetyl glutamic acid is a positive allosteric stimulator of CPS-1.

The second enzyme in the pathway, ornithine transcarbamoylase transfers the carbamoyl phosphate to ornithine to generate citrulline.

The third enzyme is located in the cytoplasm, while the first two enzymes are in the mitochondria. The argininosuccinate synthase next adds the amino acid aspartate to citulline to make argininosuccinate requiring the breakdown of one ATP to AMP and pyrophosphate, which is equivalent to the use of 2ATPs in the reaction.

In the fourth step, the argininosuccinate is lysed by argininosuccinate lyase forming arginine, an amino acid, and fumarate, a TCA cycle intermediate. The fumarate can escape into mitochondria and can serially become malate and oxaloacetate. The latter can be transaminated to aspartate which can subsequently enter urea cycle, thereby illustrating the connection between the urea and TCA cycles.

The ultimate step in the urea synthesis is brought about by arginase enzyme which hydrolyses arginine into urea and ornithine. The ornithine can enter the mitochondria to bind another mole of carbamoyl phosphate and initiate another cycle of reactions, pointing towards the catalytic requirement of ornithine, analogous to the role of oxaloacetate in the TCA cycle.

The energy requirement of the formation of one mole of urea is equivalent to 4ATP molecules, but if we consider the fumerate going to the TCA cycle, forming malate and then oxaloacetate, there is one NADH produced in the malate dehydrogenase step and this NADH can synthesise 2.5 ATPs. So, the net requirement can be taken as 1.5 ATPs.

During starvation, the catabolism of glycogen, TAG and proteins increase. The increased degradation of proteins leads to more production of ammonia which is required to be detoxified to urea, thereby necessitating elevated activities of urea cycle enzymes.

There is carbamoyl phosphate synthase-2, located in the cytoplasm, which is functioning in the synthesis of pyrimidine nucleotide synthesis, whereas CPS-1, associated with urea synthesis is present in the mitochondria.

Abnormalities in the formation of urea

As the ammonia is detoxified to urea any defect in the pathway owing to defective enzymes results in ammonia intoxication known as hyperammonaemia. The deficiency of various enzymes of the urea cycle CPS-1, OTC, argininosuccinate synthase, argininosuccinate lyase and arginase may be seen. It can lead to mental retardation, vomiting and alkalosis.

The deficiency of the first two enzymes, carbamoyl phosphate synthase-1 and ornithine transcarbamoylase cause more serious effect because of severe hyperammonaemia. The ammonia may combine with alpha-ketoglutarate, depleting the intermediates to TCA cycle, thereby decreasing the energy generation due to a slow TCA cycle. Another effect may be due to the formation of glutamine from glutamic acid and ammonia which may deplete glutamate. As gamma amino butyric acid (GABA), a negative neurotransmitter, is formed by decarboxylation of glutamate, a decreased concentration of glutamate reduces the synthesis of GABA.

Hepatic coma can result from diseases of the liver. Hyperammonaemia occurs invariably in hepatic coma. There will be CNS dysfunction and convulsions.

Blood urea level

Blood urea level is done to assess the efficiency of kidneys. The normal level of blood urea is 20-45mg/dl. Increased blood concentration of urea is called as uraemia which indicates kidney dysfunction. Another indicator of kidney dysfunction is creatinine.

One carbon metabolism

One carbon metabolism is associated with folic acid and mention of this has been done with the discussion on folic acid. The one-carbon groups are contributed by amino acids like serine and glycine (producing methylene THFA), histidine (contributing formimino THFA) and Tryptophan (forming fomyl THFA). The other one-carbon units include methenyl, hydroxy methyl and methyl.

All these one-carbon groups can be converted between them except methyl THFA which can give its methyl group only to vitamin B_{12} as explained under the section on water soluble vitamins.

These one-carbon units are specifically required in various synthetic steps like C_2 and C_8 of purines, methylation step in epinephrine synthesis and the synthesis of TMP from UMP.

Metabolism of sulphur containing amino acids

The sulphur amino acids are methionine and cysteine.

Methionine

The active form of methionine is S-adenosyl methionine (SAM) which is formed by reaction between methionine and ATP. The adenosyl group of ATP is transferred to methionine and triphosphate is split into pyrophosphate and phosphate. The adenosyl methionine can transfer its methyl group to several methyl group acceptors to form methylated products like creatine, choline and epinephrine. Methionine is an essential amino acid.

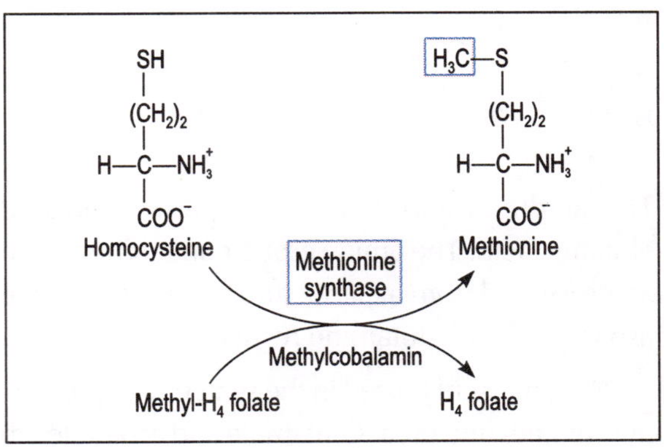

Fig. 17.3: Synthesis of methionine

Cysteine and glutathione

Cysteine helps in the stabilisation of polypeptide structure by forming interchain and intrachain disulphide bonds to form cystine. It is required for glutathione synthesis. The glutathione is gamma glutamyl cysteinyl glycine. The glutathione is important in the absorption of amino acids in addition to its role in the maintenance of RBC membrane and prevention of haemolysis. Along with glutathione, glutathione peroxidase, glutathione reductase, NADPH and selenium are also involved in the detoxification of hydrogen peroxide which otherwise can cause haemolytic crisis.

Glutathione is necessary for the other conjugation reactions and also for reducing met-haemoglobin to haemoglobin.

The inborn error of amino acid metabolism of cysteine is cystinuria due to a defective absorption of cysteine and lysine. Cystinosis is another abnormality associated with cysteine.

Homocysteinuria

Homocysteinuria is another inborn error of amino acid metabolism. Elevated levels of homocysteine can result from the deficiency of vitamin B_{12} and pyridoxine. Raised concentrations of homo-cysteine are associated with coronary artery disease.

Acidic amino acids

The acidic amino acids are aspartic acid and glutamic acid. The amide group containing amino acids formed from aspartic and glutamic acid are asparagine and glutamine respectively.

Aspartic acid is used in the synthesis of purines, pyrimidines and urea. Glutamate and aspartate are used as neurotransmitters. This stimulates nitric oxide synthase which forms nitric oxide which can bring about neuronal excitation.

GABA, as mentioned earlier, produced from glutamate is a negative neurotransmitter.

Glutamine is required for the synthesis of purines, pyrimidines, amino sugars and nicotinamide.

Asparaginase is an anticancer agent because it will deprive the cancer cells of the amino acid asparagines, as the cancer cells are unable to synthesise asparagine.

Basic amino acids

The basic amino acids are lysine arginine.

Lysine is an essential amino acid which is lacking in cereals. It is necessary for the synthesis of carnitine which is involved in the transport of fatty acyl CoA from cytosol to mitochondria. Lysine and hydroxy lysine are crucial for the cross-linking in collagen.

Arginine is used for the formation of creatine to make creatine phosphate, necessary as an energy provider for muscles. Arginine is converted to nitric oxide (NO) which has important functions.

Nitric oxide

It is classified as reactive oxygen species and is formed from arginine with the help of the enzyme nitric oxide synthase which converts arginine into citrulline and NO. The enzyme needs NADPH and molecular oxygen.

As NO is a vasodilator, it controls blood pressure. A deficiency can lead to hypertension. NO functions as a neurotransmitter and also has bactericidal action. It inhibits platelet aggregation and also relaxes smooth muscles.

Nitroglycerine and nitroprusside form NO which dilates coronary arteries and is employed for treating angina pectoris.

NO increases the blood flow into the penis, helping in its erection, making it useful in the treatment of impotence.

Branched chain amino acids

They are valine, leucine and isoleucine. Leucine is ketogenic amino acid. All these three amino acids are metabolised by reactions bearing resemblance.

Maple syrup urine disease is an inborn error of branched chain amino acids due to the deficiency of decarboxylase enzyme leading to the accumulation of the corresponding amino acids.

Metabolism of phenyl alanine and tyrosine

Phenyl alanine, an essential amino acid, can be converted to tyrosine by an irreversible enzyme, phenyl alanine hydroxylase. It is not possible to form phenyl alanine from tyrosine. A deficiency of phenyl alanine hydroxylase causes phenylketonuria.

In the step 2 tyrosine transaminase converts tyrosine to parahydroxyphenyl pyruvate in presence of pyridoxal phosphate. The third step is the formation of homogentisic acid from parahydroxyphenyl pyruvate, catalysed by parahydroxyphenyl pyruvate hydroxylase, requiring copper and ascorbic acid.

In the fourth step, the homogentisic acid is transformed into maleyl acetoacetate. The enzyme partaking is homogentisate oxidase. The fifth step is one of isomerisation, wherein maleylacetoacetate is isomerised to fumaryl acetoacetate by a specific isomerase.

The next step is of hydrolysis of fumaryl acetoacetate to acetoacetate and acetoacetate.

Phenylketonuria

This is the most commonly occurring inborn error of amino acid metabolism.

As stated, the phenylketonuria is as a result of the absence of the enzyme phenylalanine hydroxylase which is responsible for the synthesis of tyrosine from phenylalanine. There may be a deficiency of tetrahydrobiopterine also. This leads to the accumulation of phenylalanine in the blood. The excess phenylalanine is converted to phenyl

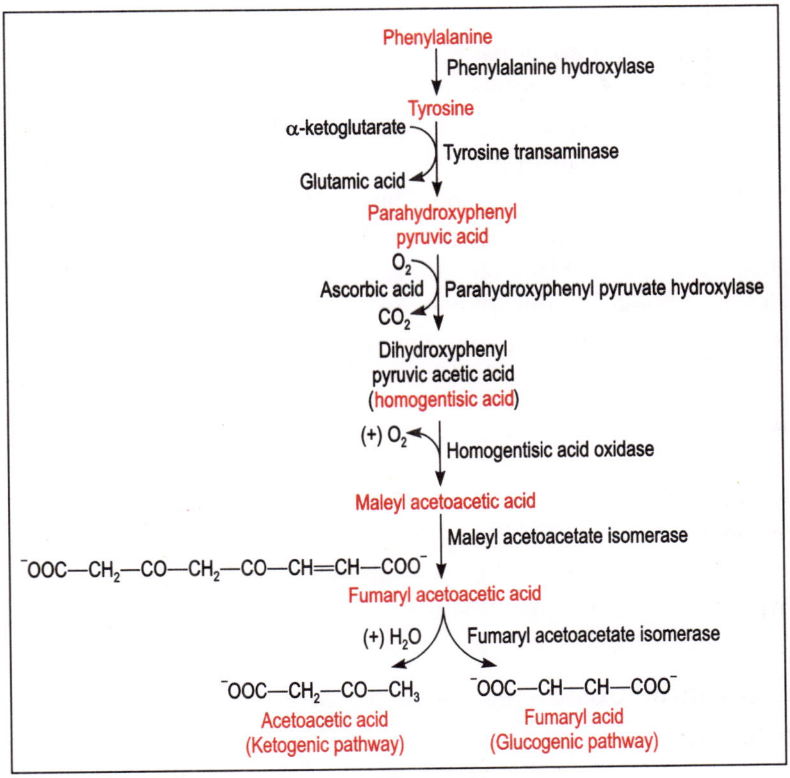

Fig. 17.4: Catabolism of phenylalanine and tyrosine

pyruvate, phenyl acetate and phenyl lactate. The child becomes mentally retarded with a low IQ and exhibits seizures and agitation. They are fair complexioned due to less pigmentation as tyrosine is necessary for melanin formation and the affected children emit a mousy odour.

The diagnosis can be made by a high blood level of phenylalanine and also by using Southern blot technique.

The treatment consists of feeding the baby with a diet having least amounts of phenylalanine. A tapioca based diet could suffice, continuing such a diet for a period of 6 years, when the brain becomes fully developed. The suspected mother's amniotic fluid screening can be done for the abnormality, and if found positive, the mother also should be advised to take a phenylalanine deficient diet during the tenure of conception.

Alkaptonuria

The biochemical abnormality in alkaptonuria is the deficiency of homogentisate oxidase enzyme which converts homogentisate to maleyl acetoacetate, resulting in the excretion of homogentisic acid in the urine. This is a benign condition and does not affect normal life. The urine after voiding turns black due to the formation of alkaptone bodies. The alkaptone may be deposited in body tissues and is called as *ochronosis*.

Albinism

Albinism results when the three is no synthesis of melanin due to the lack of tyrosinase enzyme. The skin, hair and eye lashes are all white or hypopigmented. The skin is sensitive to sunlight.

Important substances formed from tyrosine

The thyroid hormones are important substances synthesised from tyrosine by iodination. They are triiodothyronine (t_3) and tetraiodothyronine (t_4). The T_3 and T_4 have a profound effect on the functions of the body. The skin and hair pigment melanin is also formed from tyrosine.

The dopamine, epinephrine and norepinephrine are made from tyrosine. They are formed in the adrenal medulla.

The synthesis of catecholamines begins with the synthesis of dihydroxyphenylalanine (DOPA) by the enzyme tyrosine hydroxilase acting on tyrosine. In the next step, the DOPA decarboxylase enzyme in presence of pyridoxal phosphate decarboxylates DOPA to the corresponding amine, the dihydroxyphenylamine (dopamine). The dopamine is further hydroxylated to norepinephrine (noradrenaline), the enzyme catalysing the reaction being dopamine hydroxylase. The norepinephrine (noradrenaline) is methylated to epinephrine (adrenaline) by methyl transferase, the S-adenosyl methionine acting as the methyl donor.

The adrenaline and noradrenaline cause lipolysis and glycogenolysis. They also increase blood pressure.

Tryptophan metabolism

Tryptophan is a heterocyclic amino acid and it can be converted to niacin. Another product of tryptophan is serotonin (5-hydroxy tryptamine). It is a "well-being" hormone and it induces sleep. Melatonin is another compound produced from tryptophan. It is associated with sleep-wake cycles.

Hartnup's disease

Tryptophan is excreted in the urine in this disease. The absorption of tryptophan from the intestine is also defective. Pellagra is seen due to low levels of tryptophan as niacin is not formed.

Special products formed from amino acids

Several important substances are derived from amino acids other than proteins.

These consist of haem which requires glycine for its synthesis in the initial step of ALA synthase in which glycine and succinyl CoA condenses.

Another substance is glutathione requiring glutamate, cysteine and glycine for its formation. Glutamate is converted to GABA. Cysteine is necessary for the synthesis of taurine, which is required for the synthesis of tauro cholic acid—a bile acid.

Creatine and creatine phosphate need arginine and methionine for their formation. The active form of methionine, S-adenosyl methionine, is needed in methyl transfer reactions like formation of thymine, epinephrine and choline. Methionine is also the first amino acid to be incorporated into a protein. Carnitine, the transporter of acyl CoA from cytosol to mitochondria for the purpose of beta oxidation, is formed from lysine and methionine.

Several important products are obtained from the aromatic amino acids phenyl alanine and tyrosine like dopamine, norepinephrine, epinephrine, thyroxine and melanine and tyramine.

The heterocyclic ring containing amino acid tryptophan can give rise to the vitamin niacin, the neurotransmitter serotonin and a hormone melatonin. Tryptamine is formed when tryptophan is decarboxylated.

Histidine is decarboxylated to histamine, the mediator of allergic response.

The one-carbon units, carried by THFA, like formimino, formyl, hydroxyl methyl, methyl, methylene and methenyl are all derived from different amino acids.

chapter 18

DNA Replication

The DNA replication is the process of synthesis of a new strand of DNA complimentary to the parent strand and, thereby forms a double stranded DNA molecule carrying identical information of the parent DNA molecule and this process takes place during cell division. The whole DNA of the nucleus has to be replicated with utmost accuracy and fidelity. To ensure that the process is foolproof, great care is observed at the time of addition of the nucleotides in the growing strand of DNA (daughter strand), the selection of nucleotides being based on the base-pairing rules, which is followed by a follow-up mechanism to find out whether any wrong nucleotides have been inserted, and if it is found to be the case, to remove the incorrect nucleotides and substitute these by the proper nucleotides by DNA damage repair enzymes. Such a double monitoring system can ward off many of the prospective mutations and their deleterious effects. To achieve such a high level of accuracy, the process comprises several proteins and enzymes.

The DNA replication is a semi-conservative process, meaning that the daughter DNA molecules shall have one strand of the parent DNA and a newly synthesised complimentary strand. The replication proceeds from 5′ to 3′ direction in the newly synthesised strand. At the same time, the template strand is read from 3′ to 5′. There should be first unwinding of the parent DNA, as it exists as a supercoiled molecule. This is required for the DNA polymerase to get proper access into the interior of the DNA molecule. There can be several origins of replication in the DNA, but they are located at different sites, sufficiently separated. The replication process starts with a short primer of RNA of about 20 to 200 nucleotides in length.

The Process of Replication

In the eukaryotes, the topoisomerases uncoil the DNA supercoil and relieves the tension in the molecule. There are topoisomerases-I and II, with the type-I making a nick in one strands of DNA and then rotating it with subsequent resealing of the nick. The type-II variety makes nicks in both the strands of DNA prior to rotating and resealing the nicks. In the prokaryotes, the gyrases serve the corresponding function.

The next set of enzymes required for replication is helicases which further uncoils the DNA and separates the two strands by breaking the hydrogen bonds. Each of these two parent strands now serves as a template for the synthesis of a complimentary daughter strand. This produces the replication fork or replication bubble.

The third enzyme needed is RNA primase which attaches a small RNA primer having a length of about 20 to 200 complimentary nucleotides to the 3′ end of the parent strand of DNA. The system also needs single strand binding proteins (SSB) which bind to the single strands of DNA (the parent strand and the growing daughter strand) so as to prevent premature reannaeling of the single strands. The new nucleotide attached to the RNA primer and the subsequent nucleotides polymerised are deoxyribonucleotides. The selection of the new nucleotide to be polymerised into the growing daughter strand depends on the type of the corresponding nucleotide, present on the parent strand. For example, if there is AMP present in the parent strand, then a TTP approaches

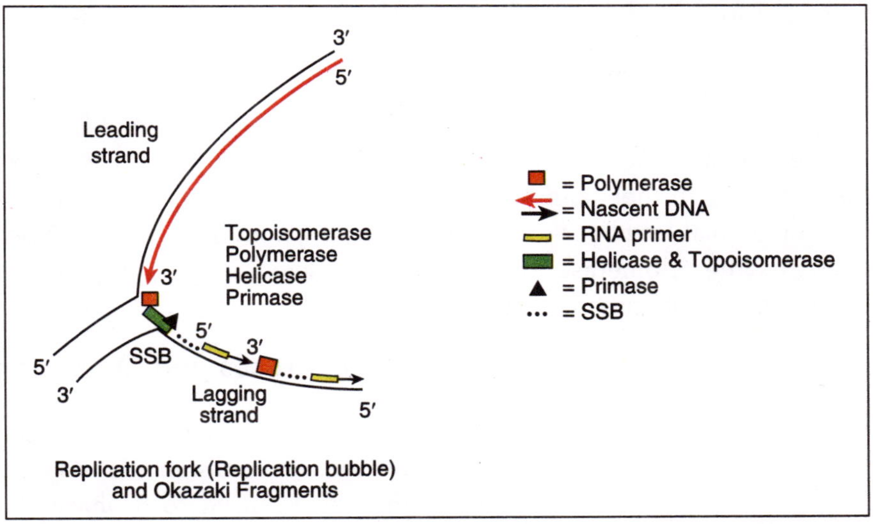

Fig. 18.1: DNA replication

and is hydrogen bonded to AMP of the parent strand, while the 3′ hydroxyl group of the recently added nucleotide on the growing daughter strand carries out a nucleophylic attack on the alpha phosphate of the incoming nucleoside triphosphate, namely, TPP, thereby incorporating it into the growing daughter strand by 3′5′ phosphodiester bond, accompanied by the liberation of a pyrophosphate. This reaction is brought about by DNA polymerase. This process gets repeated till the daughter strand completes its synthesis strictly adhering to the base pairing rule of AMP being bound to TMP, and GMP being bound to CMP. As mentioned, the daughter strand is synthesised from 5′ to 3′.

One of the two parent strands, replicated continuously and known as the leading strand is growing towards the replication fork, in the 5′ to 3′ direction. The other strand, replicated discontinuously in short spurts, designated as the lagging strand, grows from the replication fork, but the direction of growth remains the same, namely from 5′ to 3′. Therefore, the two strands are growing in opposite direction. Several Okazaki fragments, with a small piece of RNA, the primer, and a long DNA are generated on the lagging strand.

The DNA polymerase in *E. coli* was first isolated by Arthur Kornberg. In prokaryotes there are 3 DNA polymerases, namely, polymerase III is the main DNA replicating enzyme and polymerases II and I are for DNA repair. The 5 polymerases of eukaryotes are DNA polymerase epsilon (synthesis of leading strand), delta polymerase (synthesis of lagging strand), polymerase beta (DNA repair), polymerase alpha (primase) and polymerase gamma (replication of mitochondrial DNA).

Inhibitors of DNA synthesis include ciprofloxacin, novobiocin and etoposide which inhibit bacterial gyrases, and hence, are effective antibacterial agents.

The inhibitors of mammalian topoisomerases can be employed for the treatment of cancer. Examples include adriamycin and nalidixic acid.

Telomerase

The telomerases are the end portions of chromosomes which may fail to get replicated during cell division. This happens in the 3′ end portions of the parent DNA strand. After the removal of the RNA primer from the 5′ end of the daughter strand, the gap left over is not filled by properly base paired deoxyribose nucleotides. As a result of this, the parent DNA remains single stranded in its 3′ area and deoxyribonucleases degrade the single stranded portion as it can be easily hydrolysed. This further leads to the shortening of chromosomal length for each cell division. Normally, the telomerase enzyme restores the length of the chromosomes as they act as reverse transcriptase and also contain the required template. The telomerases are RNA containing enzymes and are called as *ribozymes*.

The telomerase activity goes on decreasing with age and is less in old age in whom the chromosome length is also shortened. The telomerase activity is found to be elevated in malignant state. Hence, the inhibitors of telomerase are being tried in the treatment of cancer.

Cell Cycle

The synthesis of DNA takes place only at specific time during the lifespan of the cell, known as synthetic phase of 'S' phase. This is separated from the mitotic phase 'M' phase, by non-synthetic periods known as gap 1 (G_1) and gap 2 (G_2), occurring before and after S phase respectively. The DNA synthesis occurs only once and only once in the lifespan of the cell. When once the chromatin is replicated, it is marked so as to prevent its further replication, but, the exact mechanism is under investigation. Certain regulatory proteins called as cyclins regulate the entry of the cell from one phase to another. The cyclins increase and decrease at specific times and hence, the name. The cyclins activate cyclin dependent protein kinases (CDKs) which phosphorylate other enzymes and proteins needed for the progression of the cell to the next phase. The CDK_4 and CDK_6 are nuclear proteins and are serinethreonine protein kinase. The retinoblastoma (Rb) protein regulates cell cycle by binding to transcription factor (E_2F), inactivating it. The E_2F is necessary for the progression of the cell from G_1 to S phase. As Rb protein binds to E_2F and inactivates it, the cell remains in the G_1 phase, but the phosphorylation of Rb gene by CD_4 and CD_6 causes the E_2F is released from Rb protein-E_2F complex. The E_2F can ensue the progression of the cell from G_1 phase to S phase. The Rb gene is considered as an oncosuppressor gene.

Onco-viruses and Onco-genes

The cancer causing viruses (onconviruses) and oncogenes can disrupt the control of the entry of the cell from G_1 to S phase. The excessive production cyclin and appearance cyclin at improper time can result in unrestrained cell division. The bcl oncogene, associated with B-cell lymphoma happens to be a cyclin gene. The products of several oncogenes are protein in nature and function as growth factors like platelet derived growth factors and receptors of endothelial growth factor. Some oncogenes produce protein kinases which are specific for tyrosine residues. Oncogenic viruses introduce oncogenes into the body which activates proto-oncogenes leading to malignant transformation. The P_{21}, the product of ras gene inhibits adenyl cyclase, but, the C-ras oncogene product is mutated P_{21} and hence, cannot inhibit adenyl cyclase required to mediate hormonal action.

Onco-suppressor Genes

The onco-suppressor genes prevent cancer but abnormalities in these genes may result in cancer. An example of oncosuppressor gene is Rb gene and P_{53} gene. The Rb protein and its corresponding gene has already been explained.

The P_{53} gene is so called because it encodes for a phosphoprotein with a molecular weight of 53000. The P_{53} gene blocks the DNA replication especially when the DNA is damaged. It also activates another gene, called as P_{21}, which further blocks the cell division till the damage in the DNA is repaired. If the DNA is damaged beyond repair, the cell is guided for apoptosis or programmed cell death. The P_{53} gene can combine with oncogenic viruses to suppress cancer production. The P_{53} gene is absent in cancer patients or severely damaged mutations. The single nucleotide polymorphisms can make the P_{53} gene devoid of its anticancer properties. Other tumour suppressor genes include $BRCA_1$ and $BRCA_2$, the letters standing for breast cancer. These genes prevent cancer, but mutations in these genes can cause cancer of breast and ovaries.

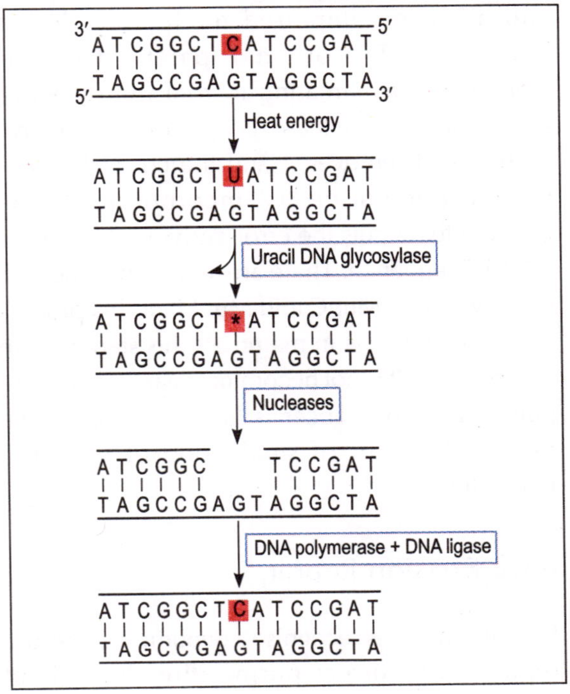

Fig. 18.2: Nucleoside excision repair

DNA Damage and Repair Mechanisms

The maintenance of the integrity of the genetic information in the form of nucleotide sequence is very much essential for every organism and also for the genetic stability of different species. As the DNA constantly has to face the damaging effects of various factors like ultraviolet rays, x-rays, gamma rays, reactive oxygen species, mutagenic chemicals, etc., the organisms should have the necessary enzymes to identify the damages in the DNA and repair the affected areas, so as to ensure their survival. Adhering to the base pairing rules during the process of replication is one step towards the maintenance of the genetic stability. This is further reinforced by a follow-up mechanism to identify any defective or mismatched nucleotide incorporation and immediately removing the improper nucleotides and substituting these with the correct ones. Hence, such proof reading does take place to maintain the genetic integrity.

The damages occurring to DNA are single-base changes, two-base changes, chain-breaks and cross-linkages.

The single-base changes include depurination (removal of purines), deamination of cytosine to uracil, deamination of adenine to hypoxanthine, alkylation of base, insertion or deletion of nucleotides and base analogue incorporation.

The two base alterations include ultraviolet induced thymine-thymine dimer.

The DNA chain-breaks can be caused by ionising radiations and reactive oxygen species.

Another type of damage is because of cross-linkage between bases and between bases and nucleoproteins.

The mechanisms of repair of damaged DNA comprises of mismatch repair, base excision repair, nucleotide excision repair and double strand break-repair.

Mismatch Repair

It is useful for correcting single base change or a defective short segment of 2 to 5 unpaired loops which may result due to a slip by DNA polymerase and adds few unpaired nucleotide during replication. The original (parent strand) is methylated demarcating it from the defective strand. An endonuclease recognises the defective portion and removes it and the correctly base-paired nucleotide is incorporated. The endonuclease cuts the error-carrying strand near a GATC sequence. There are 3 proteins required, namely Mut S, Mut C and Mut H for this process in *E. coli*, whereas in mammals this step is more complicated. It is not becoming clear that the non-polyposis colon cancer is linked to a defective mismatch repair of small loop of unpaired nucleotides.

Base Excision Repair

Specific enzymes recognise a depurinated site and replace the correct purine directly without disturbing the phosphodiester bond. Cytosine, adenine and guanine are spontaneously deaminated to uracil, hypoxanthine and xanthine respectively. As these bases are a non-entity in the DNA, the N-glycosylases recognise these abnormal bases and removes these bases. An endonuclease cuts the DNA on both sides of the defect and removes a few bases. Subsequently, a repair DNA polymerase adds the proper nucleotides in the vacant space with the ligase rejoining the DNA. Thus, the base excision repair can effectively correct a single base change but is rather not useful since long stretches of DNA are damaged.

Nucleotide Excision Repair

For repairing longer regions of DNA (about 30 bases in length) the nucleotide excision method is employed using exinucleases. Such changes are caused by UV light induced thymine-thymine dimer formation, smoking, ionising radiation, cancer chemotherapeutic agents and a variety of chemicals. These agents can cause cross linkages, chain breaks and base alterations. The exinucleases remove a region of about 30 odd nucleotides in length and this vacant portion is filled by proper nucleotides and then joining the DNA by ligase enzyme.

Xeroderma pigmentosum is an autosomal recessive genetic disease and the patients with this disorder exhibit marked sensitivity to UV light and develop skin cancers. These patients have defective repair of thymine-thymine dimers.

Some other diseases associated with defective DNA damage repair mechanism are Fanconi's anaemia and ataxia telangiectasia, characterised by lymphoreticular neoplasm and ataxia.

The double strand break repair is similar to immunoglobulin rearrangement occurring physiologically. The chain-break repair involves 2 proteins, ku and a DNA dependent protein kinase.

Some enzymes of DNA repair are also involved in transcription and cell cycle (DNA synthesis). They are also linked to normal gene rearrangement.

chapter 19

Transcription and Post-Transcriptional Modifications

Transcription is the process by which RNA is transcribed from the template strand of DNA. The synthesis of the new strand of RNA occurs from 5′ to 3′, in a way similar to that of replication. Meanwhile, the template strand is read from 3′ to 5′, that is, in an antiparallel manner. The enzyme required for transcription is DNA dependent of RNA polymerase. Roger Kornberg got Nobel Prize in 2006 for his work on transcription, whereas, Arthur Kornberg was awarded Nobel Prize in 1959 for his work on DNA replication.

There are 3RNA polymerases in mammals. The RNA polymerase type I or A synthesises rRNA, type III or C is responsible for the synthesis of tRNAs and type II or B is the major enzyme synthesising mRNAs and SnRNA. It has 2 large and 12 smaller subunits. The phosphorylated form of RNA polymerase II is active.

The bacterial RNA polymerase has 2 alpha and 2 beta subunits in addition to one sigma factor, one omega unit and 2 zinc molecules. The sigma factor is for the recognition of the promoter site on the DNA.

The transcription process consists of 4 steps, namely, initiation, elongation, termination and post-transcriptional modifications.

The sequence of nucleotides in the RNA is complementary to one of the two strands of DNA, known as template strand, the other strand is called as the *antitemplate strand*, also called as *coding strand* of that gene. It is so called because it is exactly similar to the mRNA with the exception of UMP replacing TMP, in the RNA. Further, the template strand of all the genes will not be

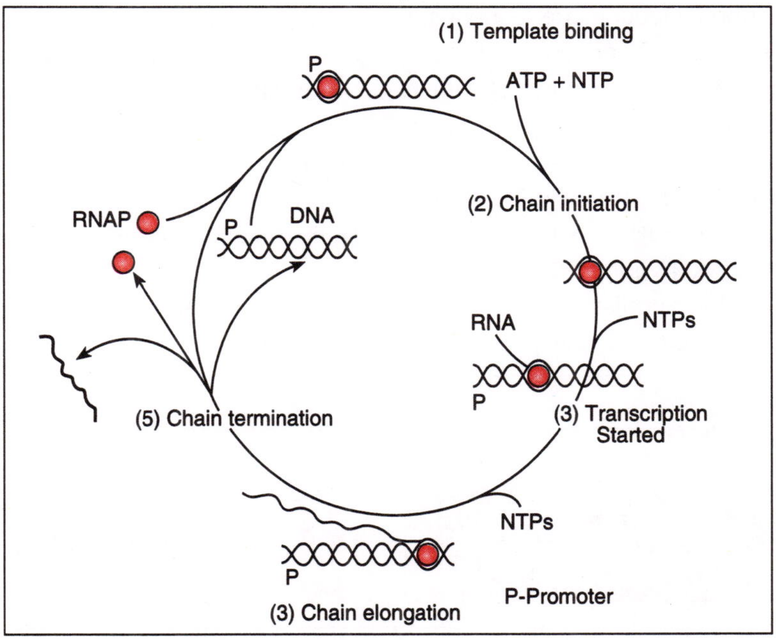

Fig. 19.1: Transcription: Formation of RNA

necessarily the same strand of the double helix. Thus, a particular strand of a double helix will serve as the template strand of some genes and the coding strand of other genes.

Initiation of transcription

The DNA dependent RNA polymerase attaches at specific site, the promoter region on the template strand. In prokaryotes the RNA polymerase combines with the sigma factor to help in the recognition of the promoter site. The first ribonucleotide transcribed is a purine (A or G). After the initiation, the sigma factor is released from the holoenzyme.

Elongation

The elongation of the RNA takes place by the addition of ribonucleotides, catalysed by RNA polymerase, as dictated by the sequence of deoxyribonucleotides present in the template strand of DNA and as interpreted by the base pairing rules. The ribonucleotides are sequentially added to the 3′ hydroxyl group of the last nucleotide of the growing RNA transcript with its hydroxyl group carrying out a nucleophilic attack on the alpha phosphate of the incoming ribonucleotide triphosphate forming a 3′5′ phosphodiester bond with the liberation of pyrophosphate. As the RNA polymerase progresses along the length of the template strand of DNA molecule, DNA unwinding is essential so as to provide access to appropriate base pairing to the corresponding nucleotides of the coding strand. The transcription bubble (the unwound region of DNA) is about 20 base-pairs long/polymerase molecule. The RNA polymerase has an inbuilt unwindase activity to open the DNA helix. This process of elongation continues till the gene is fully transcribed into the corresponding Hn-RNA, consisting of both introns and exons.

Termination

The specific termination signals located on the template strand of DNA are recognised by the rho protein factor which separates the DNA-RNA complex. This spells the end of the process of transcription. This primary transcript has to be further processed to form mature mRNA.

Promoters and transcription start signals on DNA

The bacterial promoters are relatively simple and span about 45 bases on the template strand of DNA. About 35 base-pair upstream from the start site and there is a sequence of TGTTGACA in the 5′ to 3′ direction to which RNA polymerase binds. There is another sequence of TATAAT, about 10 nucleotides upstream again in the 5′ to 3′ direction on the template strand and is referred to as TATA box. These two signals specify where the transcription should commence or which is the start site.

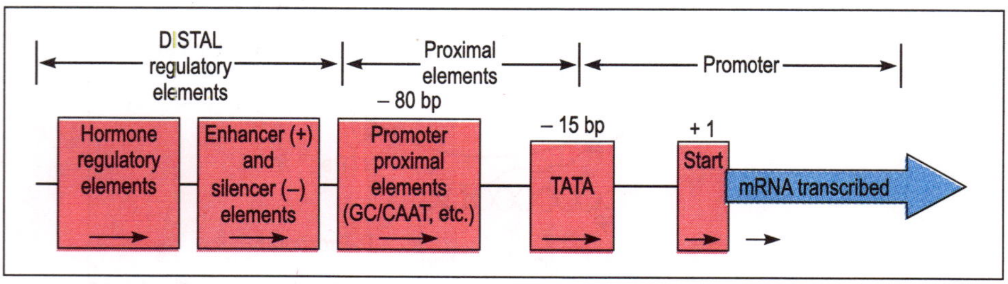

Fig. 19.2: Regulation of transcription

The transcription start signals are expectedly more complex in eukaryotes which have the "where" (start site) and "how frequently" signals. The latter specify the frequency of transcription. About 32 nucleotides upstream from the start site in the eukaryotes, there being a sequence of TATAAAAG which is the counterpart of TATA box of prokaryotes. The TATA box determines the transcription start site. The transcription requires the bind of TATA-binding protein to TATA box. Further, there are sequences known as GC and CAAT boxes which specify the frequency of transcription after being bound to specific proteins. Mutation in the above cited sequences can reduce transcription frequency. The eukaryotes have another set of sequences known as enhancers and silencers (repressors). The enhancers increase and silencers decrease the frequency of transcription. The eukaryotes are also equipped with hormone response elements (HRE) on the DNA for binding to hormones which regulate transcription. The DNA dependent RNA polymerase II is active when phosphorylated by kinases. There are several activators and coregulators for transcription. All these make the eukaryotic transcription process more complex.

Posttranscriptional modification

In prokaryotes there is very little post-transcriptional modification taking place and hence, the mRNA transcript undergoes translation prior to the transcription being complete.

In eukaryotes, on the other hand, the transcription and translation are completely separated, the former being a nuclear event and the latter being a cytoplasmic one. Moreover, the primary transcript (Hn-RNA) in eukaryotes has to undergo extensive processing before enacting its role in protein biosynthesis. The modifications occurring prior to the maturation of mRNA include capping at the 5′ and with 7-methyl GTP, attachment of polyadenosyl (A) tail at the 3′ end and splicing consisting of removal of introns (non-coding sequences) and joining of exons (exons).

The attachment of 7-methyl GTP cap at the 5′ end of mRNA helps in its recognition, as all the other nucleic acids are devoid of nucleotide triphosphate. The attachment of poly A tail at the 3′ end helps in the stability of mRNA.

The removal of non-coding introns and the joining of coding exons are termed as splicing. The formation of spliceosomes is a prerequisite for this process which ensures the proper recognition of the intron-exon junction point as improper removal of splicing can result in diseases like thalassaemias. The generation of antibodies to spliceosomes causes systemic lupus erythematoses (SLE). The splicing is catalysed by ribozymes containing RNAs like U_1, U_2, U_4, U_5 and U_6, so named because of their richness in uracil. The complex of all the five RNAs (Sn RNAs) bind with proteins to form small nuclear ribonucleoproteins, also termed as "snurps". It is perceived that the RNAs were functioning as catalytic agents in the long, long past, before the proteins took over this function and hence, the ribozymes can be considered vestigial molecules.

Messenger RNA editing

Certain mRNAs are subjected to editing in different tissues. For example, the mRNA to be translated to apolipoprotein-B is edited in intestinal mucosal cell and forms the apo-B-48, the apoprotein of chylomicrons, formed in the intestinal mucosal cell. The VLDL and IDL contain the apo-B-100, without any editing process and this happens in the liver. In short, we can say that apo-B-48 consists of the first 48 per cent of the sequence of amino acids from the amino terminus of the apo-B-100. This can be attained by terminating the translation process prematurely in the intestinal mucosal cell, after translating 48 per cent of the apo-B-100 mRNA, whereas in the hepatic tissue, the mRNA is fully translated into protein, thereby forming apo-B-100. The molecular basis for the editing of mRNA in the intestinal mucosal cell is due to the action of a cytidine deaminase converting the codon CAA in the mRNA to UAA. As a result of this, the glutamine codon (CAA) becomes a termination signal.

chapter 20

Translation—Protein Synthesis

The process of transcription is a nuclear event while translation takes place in the cytoplasm on the ribosomes. For this the mature mRNA has to be transported to the ribosomes from the nucleus. The translation of mRNAs meant intracellular proteins occurs on the free ribosomes, while those for cell membrane proteins and for those proteins to be secreted out are formed on polyribosomes, which represent ribosomes attached to endoplasmic reticulum, known as rough surface endoplasmic reticulum, different from the smooth surface endoplasmic reticulum.

The translation requires the presence of all the 20 amino acids in the right proportion and all of which need to be activated to form corresponding aminoacyl tRNAs. There is a specific tRNA and a specific aminoacyl tRNA synthase enzyme for each amino acid. As stated earlier, the mRNAs contain codons and the tRNAs have complimentary anticodons both of which interact during the fabrication of proteins, the amino acids being incorporated into the growing protein chain, ad dictated by the codons present in the mRNA.

The translation can be divided into three steps: initiation, elongation and termination, in a way similar to transcription.

Initiation

The transcription commences from the 5′ end of the mRNA and proceeds to the 3′ end, with the 5′ end forming the amino terminus and the 3′ end forming the carboxy terminus of the protein. The first amino acid incorporated into a protein is always methionine (eukaryotes) or formyl methionine (prokaryotes). The steps of translation in eukaryotes and prokaryotes are different in many ways.

In eukaryotes there are 9 initiation factors (eIF), whereas there are only 3 eIFs in prokaryotes. The prokaryotic ribosomes are smaller (60S) than the eukaryotic ones (80S). What follows is a description of the process in eukaryotes where initiation factor, eIF_2 interacts with GTP which in turn combines with methionyl tRNA to form a ternary complex. Meanwhile, 40S ribosomal subunit combines with eIF_1, eIF_{1A}, eIF_3 and eIF_5 to form a complex. The above two complexes combine together to generate a preinitiation complex.

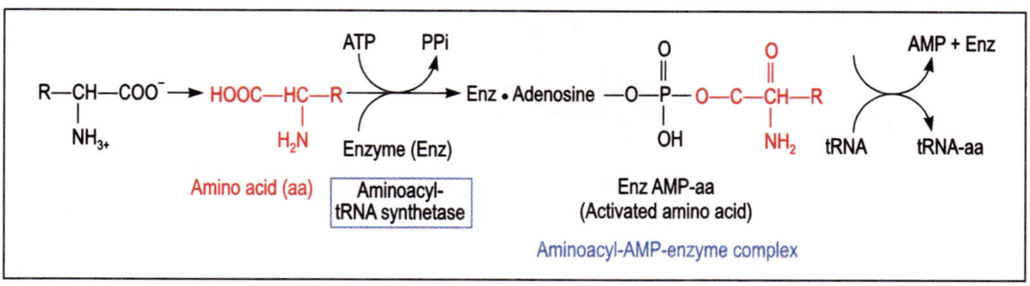

Fig. 20.1: Activation of amino acid

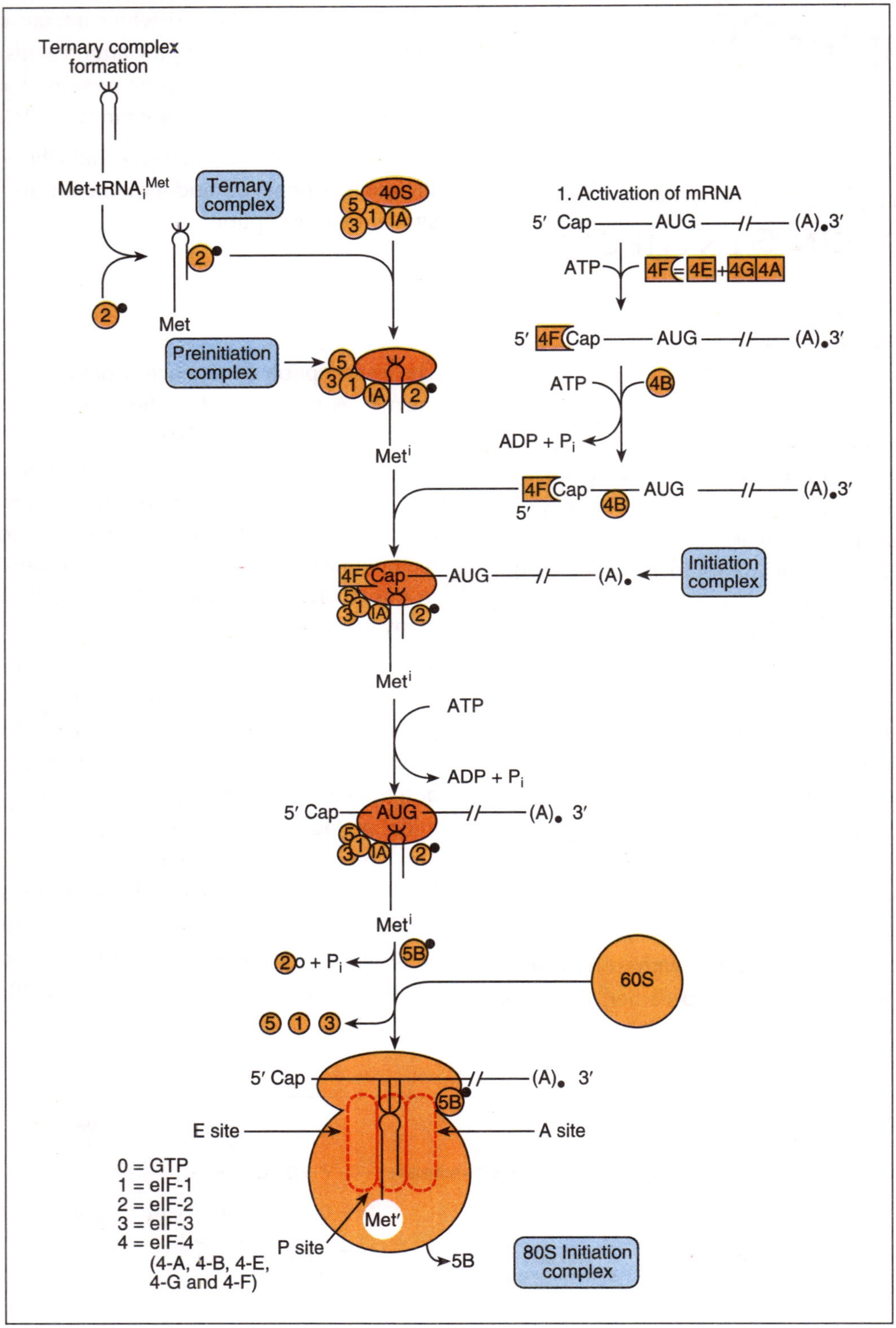

Fig. 20.2: Initiation of translation

The mRNA is next activated with ATP and the initiation factors, eIF_{4A}, eIF_{4E}, eIF_{4F} and eIF_{4G} are also attached. Next ATP is hydrolysed and initiation factors eIF_{4B} and eIF_{4F} are also bound.

The complex next combines with 43S preinitiation complex to form 48S initiation complex consisting of 40S ribosomes, mRNA, methionyl tRNA, a molecule of GTP and the initiation factors, as stated before. Now the AUG codon present in the 5′ end of the mRNA is located and the methionyl tRNA is attached by its anticodons to the AUG code.

The 60S ribosomal unit is then combined to the 48S initiation complex and forms the 80S ribosome or the whole ribosome. During the formation of 80S complex ATP is hydrolysed. This also needs eIF_{5B} and all the other initiation factors are liberated from the complex.

The whole ribosome has two sites: the aminoacyl or "A" site and peptidyl or "P" site. The methyionyl tRNA occupies the P site. The new incoming aminoacyl tRNA attached at the A site.

Elongation

As indicated before, AUG is the first codon translated and the amino acid incorporated is methionine. Now the second code on the mRNA is deciphered and the amino acyl tRNA, indicated by the code, binds an elongation factor, eEF_1 and

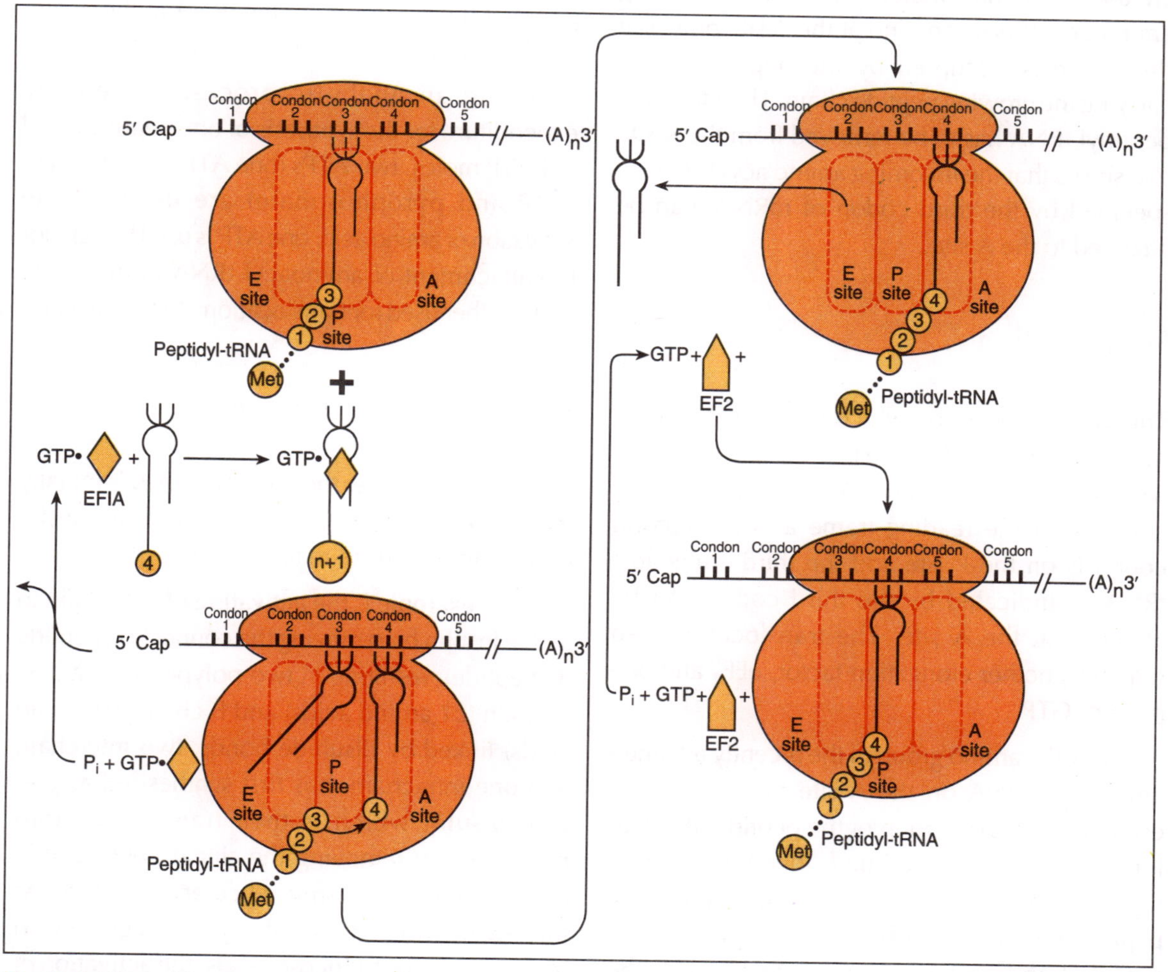

Fig. 20.3: Elongation step in translation

a GTP molecule, and this complex attaches to the A site of the 80S ribosome. With this binding the GTP is hydrolysed. There will be reaction between the amino group of the amino acid at the A site and the esterified carboxyl group of methionine to form a peptide bond and this reaction is catalysed by peptidyl transferase, which is a ribozyme, meaning an RNA demonstrating enzyme activity. This is a component of 28S RNA of 60S ribosomal subunit. As the amino acids are already activated, there is no need of extra energy for the peptide bond synthesis. With this, the growing peptide chain is attached to the tRNA attached at the A site. Meanwhile, the tRNA at the P site is free, and hence, vacates the P site to attach to another methionine molecule. The new incoming amino-acyl tRNA can bind at the A site only, but the A site is occupied by the peptidyl tRNA carrying the growing peptide chain. Therefore, the peptidyl tRNA has to be removed from the A site to P site so that the incoming amino acyl tRNA, as specified by the third codon of mRNA, can be attached to the A site.

Translocation

This is the process by which the ribosome over mRNA by a codon distance in the 5′ to 3′ direction so that the initiator codon (AUG for methionine) goes out of the reading frame and the second codon is on the P site and the third aminoacyl tRNA as indicated by the third codon can be attached at the A site. The translocation step requires another elongation factor, eEF_2 and one mole of GTP.

Now the amino group of the recently attached aminoacyl tRNA on the A site reacts with the esterified carboxyl group of the peptidyl tRNA to form a second peptide bond, thereby making a tripeptide, the reaction being brought about by peptidyl transferase, a ribozyme. The process gets repeated and protein chain elongates with the addition of amino acids as dictated by the codons of mRNA.

Termination

Any one of the three termination codons appears at the A site, after the designated number of aminoacids in the specified sequence are polymerised. There is no aminoacyl tRNA corresponding to the termination and hence, releasing factors attach to the A site in presence of GTP molecule which is hydrolysed to GDP and phosphate. The releasing factors hydrolyse the ester bond between the carboxyl group of the carboxy terminal amino acid and the hydroxyl group of the adenosine of the tRNA at the A site, resulting in the release of protein chain.

Being an anabolic process the energy expenditure for the synthesis of one peptide bond is 4 ATP moles, two ATPs (one ATP hydrolysed to AMP and pyrophosphate) are used for the activation of amino acid, one ATP is used by eEF_1 for the attachment of aminoacyl tRNA to the A site and another used for translocation along with eEF_2.

Post-translational modifications

On several instances the protein formed is initially inactive and has to be subjected to post-translational alternations to make it active.

One example is the activation of proinsulin to active insulin by removing the connecting peptide (C-peptide), resulting in two polypeptide chains, A chain (21 amino acids) and B chain (30 amino acids) linked by disulfide bonds, two interchain and one intrachain. It is first synthesised as pre-pro- insulin which is then transformed into proinsulin by removing the signal peptide after entering the endoplasmic reticular membrane. As the secretion of insulin should be regulated in response of blood glucose levels, the activation of proinsulin to insulin in response to hyperglycaemia is of importance.

Another example is that of the maturation of procollagen which consists of triple helix polypeptide chains, twisted one over the other. The fibroblasts synthesise the tropocollagen consisting of three polypeptide chains, each with 1000 amino acids. The tropocollagen is secreted out into the extracellular matrix, where specific peptidases cleave some of the amino acids from the amino and carboxy termini. Subsequently, the lysyl and prolyl hydroxylases add hydroxyl groups to lysine and proline to form hydroxyl lysine and hydroxy proline respectively. These reactions need iron in the ferrous state as a cofactor, which is made possible by ascorbic acid, the deficiency of which leads to scurvy due to defective collagen.

A third example of post-translational modification is the activation of enzymes of protein digestion like pepsin, trypsin and chymotrypsin which are generated from the inactive enzymes pepsinogen, trypsinogen and chymotrypsinogen, respectively by the removal of amino acids from the aminoterminus of the inactive enzymes. These processes are helpful in having the presence of these active enzymes only when the food is present in the intestine and prevent the autodigestion of the intestinal mucosa.

The activation of blood clotting factors, V, VII, IX, X and XI, which are secreted as inactive proteins, require vitamin K acting as a coenzyme for the addition of carboxyl group to the gamma carbon of glutamic acid residues of the clotting factors. Now the divalent calcium necessary for initiating blood clotting can combine with the two carboxylic groups of gamma carboxy-glutamic acid. By this mechanism clotting factors can be made available as and when required in good quantities during emergency.

The osteocalcin, an important protein seen in the bones, also requires vitamin K for the gamma carboxylation glutamic acid, which helps in the bone mineralisation with calcium deposition showing an additional role of vitamin K other than in blood coagulation.

Protein folding and chaperones

After the translation and post-translational changes the proteins have to be properly folded into 3-dimensional structures as incorrectly folded proteins are not functional.

The chaperon proteins bring about the proper folding of the proteins. Mutations in the chaperones can cause mistakes in attaining the active conformation of proteins. The functioning of the chaperons are further regulated by chaperonins. The incorrectly folded, and hence, inactive proteins are degraded mediated by a ubiquitin involving mechanism.

Protein sorting

The properly folded proteins have to be sent to correct locations, meaning nucleus, mitochondria, lysosomes, peroxisomes, etc. Specific sequences located near the carboxy terminus of the protein determine its destination. Hence, the amino acid sequences near the c-terminus are crucial in the sorting of proteins and any abnormality in this sequence can give rise to defective delivery of the proteins. For example in Zell Wager's disease the very long chain fatty acids are not oxidised as the peroxisomes do not get the enzymes for the oxidation of these fatty acids. The specific sequences like nuclear, mitochondrial, lysosomal and peroxisomal import sequences are present for transporting the designated proteins respectively to these organelles.

Inhibitors of Protein synthesis

The differences in the protein synthesis between prokaryotes and eukaryotes can be exploited to make inhibitors of prokaryotic translation, which can serve as antibiotics. Compounds like erythromycin, streptomycin, penicillin, ampicillin, tetracycline, chloramphenicol, etc. inhibit the prokaryotic process, and so, are useful antibiotics. The puromycin, an analogue of phenyl alanine, inhibits eukaryotic process and, hence, cannot be used as an antibiotic. The diphtheria toxin also inhibits eukaryotic translation.

Mutations

The mutation is a change in the sequence of nucleotides in the DNA. It may be a point mutation in which one base is replaced by another base or frame shift mutation which is because of addition of one, two or three bases or multiples of triplets.

Point mutations

The point mutations may be transition or transversion. The transition is when a purine base gets replaced by another purine base and the transversion involves the replacement of a purine by a pyrimidine or a pyrimidine by a purine. The sickle cell haemoglobin results when adenine is replaced by thymine, in the 6th code of beta chain of haemoglobin resulting in the substitution of valine in place of glutamic acid.

The point mutations can have different results like:

(a) Having no detectable effect on the product due to the degeneracy of the code. The same amino acid is incorporated if the base altered occupies the third position of triplet codon.

(b) A missense mutation takes place when the mutation leads to the replacement of one amino acid by another. One type of missense mutation may have no effect on the function of the protein product which may be the result of the replacement of one amino acid by another with similar properties; for example, the replacement of aspartic acid with glutamic acid or lysine with arginine or alanine with valine.

A missense mutation also can result in a partially acceptable protein product, an example being that of sickle cell haemoglobin which does bind and release oxygen but with less efficiency.

Another type of missense mutation can result in unacceptable protein product, an example for this being the HbN mutation of histidine replaced by tyrosine. This converts ferrous form to ferric, with the haemoglobin becoming met-haemoglobin, which cannot bind and transport oxygen, and hence, unacceptable.

(c) Another type of mutation is nonsense mutation which is due to the premature appearance of termination codons which terminates the protein synthesis before completion and produces a smaller protein with no function.

Examples of point mutations are aplenty in haemoglobin molecule. The examples include Hb Milwaukee, Hb Sydney, Hb Bristol, Hb S, Hb C, Hb D, Hb E and so on.

Frameshift mutations

These are the result of addition or deletion of varying number of nucleotides. One, two, three or multiples of three (tiplets) nucleotides may be added. The addition or deletion can lead to incorrect nucleotide codons, resulting in a garbled sequence of amino acids, after the addition or deletion, but before the addition or deletion, the amino acid sequence will be proper. Further, if an addition is followed by a deletion, or a deletion is followed by an addition, the amino acid sequence between addition and deletion shall be wrong, but before the addition and after the deletion, the amino acid sequence shall be proper.

Operon concept

The operon concept can be understood by studying the system for the metabolism of lactose by *E. coli* bacterium. When *E. coli* are grown in a medium containing both glucose and galactose, the bacteria uses only glucose as the enzymes of lactose metabolism are repressed at the gene level. The operon is a group of genes involved in the synthesis of a set of enzymes participating in a specific metabolic pathway. The Lac operon consists of three structural genes, Z, Y and A, for beta-galactosidase, permease and acetylase respectively. Adjoining these genes are an operator locus and a gene coding for a repressor molecule.

When the glucose is available in the growth medium, the repressor protein mole is expressed and is bound at the operator locus in the promoter region, thereby hindering the transcription of the Z, Y and A genes and the further synthesis of the corresponding enzymes, required for lactose metabolism. This process is called as *repression*. This shows that lactose is not oxidised by the *E. coli* bacteria for energy purpose, provided glucose is available as an energy fuel.

As glucose gets exhausted, the bacteria stops growth for some time till the lactose metabolising enzymes becomes induced or derepressed. In the discontinued supply of glucose, the lactose molecule enters into the bacteria with the help of permease enzyme and binds to the repressor molecule when it becomes weak and gets dissociated from the operator locus and the RNA polymerase now can bind at the promoter site and start to transcribe the three genes. This is known as derepression or induction. With this the bacteria are able to use lactose as an alternative source of fuel.

Such operons are also seen in eukaryotes as well.

Prions

Prions are proteinacious infectious particles which can convert a normal protein into a prion protein which in turn converts another protein abnormal and this chain reaction continues. Prions have no functions and cannot be hydrolysed. This was first seen in sheep having "scrapie" disease which was transmitted to cattles. This is known as "mad cow disease".

Alzheimer disease in humans can be similar to the prion disease where abnormal amyloid proteins accumulate in the brain forming neurofibrillay tangles affecting both sensory and motor nerves. Another prion protein involved is "tau" protein which cause memory loss. Prions possess less α-helix.

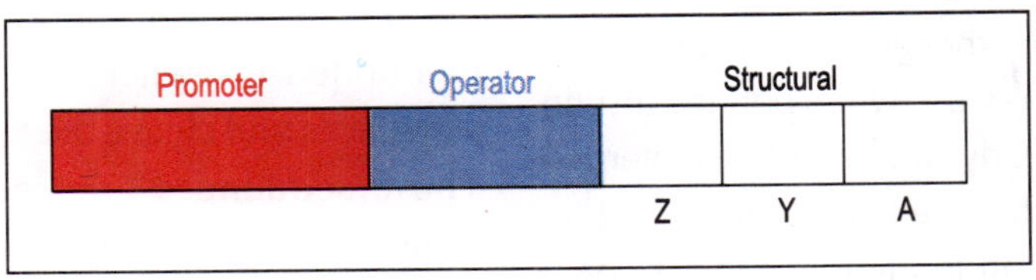

Fig. 20.4: Lac Operaton

chapter 21

DNA Recombinant Technology

This is a process by which DNA from different sources (species) are joined to create new sequences with beneficial effects. This is known as molecular cloning which is different from animal cloning, first carried out when a sheep named dolly was cloned. For example, in molecular cloning, human genes could be incorporated into *E. coli* DNA using a vector like plasmid and the bacteria starts to synthesise human proteins, for example, insulin. The insulin of other animals like ox, sheep and pig, possessing different amino acid sequences, may elicit immunologic response in the recipient over a period of use, and, so, the effectiveness is decreased. Hence, the insulin produced by genetic engineering or otherwise, called as DNA recombinant technology, could circumvent this problem of resistance due to antibody generation against insulin obtained from other animals. True that the antibody production is least, when porcine insulin is employed as it is nearly similar to its counterpart in the humans with the sole alteration present on the C-terminus (30th amino-acid) of the B chain which in humans have threonine but in pigs have alanine. Other molecules formed by this technique are growth hormone, interferons, hepatitis B vaccine, etc. But for this technique, to obtain a few units of growth hormone, several pituitary glands would have been required. In the case of hepatitis B vaccine, handling of highly infective sera from hepatitis patients would have been a must. Other applied aspects of this technology is in medicolegal cases and detection genetic abnormalities, activation of oncogenes, mutation in onco-suppressor genes and infection by bacteria and viruses and many more. This technology is useful not only for the amplification of a specific DNA molecule but also for getting the human proteins, synthesised by bacteria.

The technique requires the following tools:

(a) The target DNA which is to be recombined with the vector.

(b) Restriction endonuclease.

(c) Plasmids, bacteriophages or cosmids as vectors, depending on the size of the DNA to be recombined.

(d) *E. coli* bacterial culture

(e) Reverse transcriptase.

The plasmids are circular DNA molecules which confers antibiotic resistance on the bacteria. In genetic engineering, they are used as vector for incorporating the desired DNA into the bacterial genome. The plasmids can take up DNAs with a length up to 10 kilobase pairs. The bacteriophages incorporate larger sized DNA moles between 10-20 kilobases in length. Still larger DNAs which are more than 20 kilo-bases in size require cosmids as vectors.

The restriction endonuclease enzymes, as mentioned elsewhere, can recognise specific sequences and cleave the molecule at these sites. These sequences are known as *palindrome* sequences and they read the same in forward and backward directions. The number of REs discovered has been about 800, each of which has site specificity.

The isolation of the desired DNA (gene) to be recombined is a difficult process as the DNA is a very large molecule and to search for a particular gene and to isolate it is not an easy process. Easier

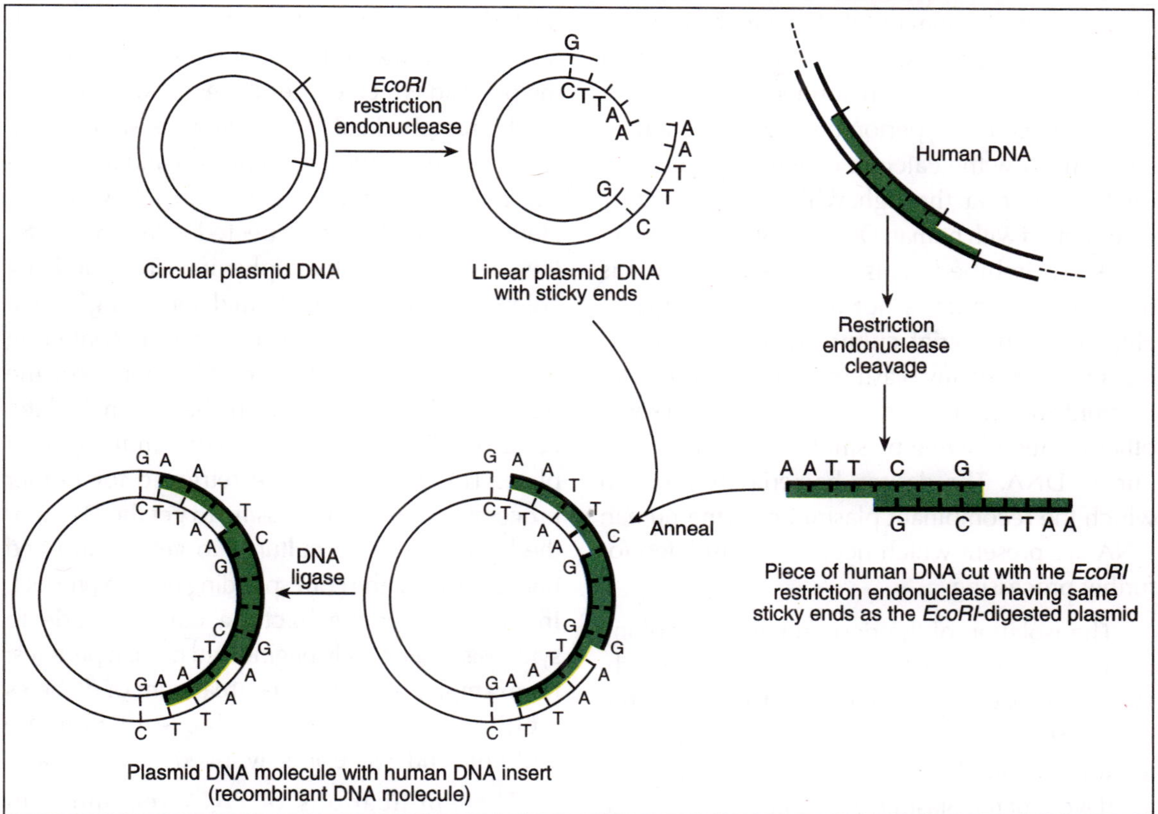

Fig. 21.1: DNA recombination of Plasmid and Human DNA

than this would be to first isolate the mRAN and then prepare the complimentary DNA from it. The isolation of mRNA is easier because of the differential expression of genes into proteins via the corresponding mRNAs. For example, insulin mRNA can be obtained from the beta cells of the islets of pancreas, albumin mRNA can be seen in the liver, growth hormone mRNA will be present in the anterior pituitary glands and so on and so forth. If we, for example, plan to isolate insulin mRNA, then the first step is to get all the RNAs extracted from the beta cells and pass the extract through a column containing poly T sequences. During the passage of the RNA extract through the column, the mRNAs, as they have the poly A tail, bind with the poly T in the column and, so, are held back. All the other types of RNAs get eluted. Later on the mRNAs are eluted from the column. From this elute the insulin mRNA can be separated.

The complimentary DNA can be prepared from the mRNA by a RNA dependent DNA polymerase, also called as *reverse transcriptase* which is seen in RNA viruses for incorporating their genetic material to that of the hosts culminating in infection. First, the insulin mRNA can be copied into a single-stranded DNA which can form a double-stranded DNA with the assistance of DNA polymerase. The DNA so formed is known as cDNA or complimentary DNA which can be incorporated into the circular plasmid to form the recombinant plasmid. For this, both the human DNA and the plasmid DNA have to be cleaved by the same restriction endonuclease so that annealing of the two can take place, forming the recombinant plasmid. This process will occur when the human DNA is incubated with plasmid. In fact, only some plasmids carry human DNA, after which these are further incubated with *E. coli* bacteria in hypertonic solution containing calcium. Now the calcium channels are opened on the bacteria through which the plasmids recombined with human DNA and without human DNA are imbibed. This process is known as transfection which is not an efficient process. Hence, the bacterial culture will contain wild bacteria without any plasmid, some bacteria with plasmid and without any human DNA, and some other bacteria having plasmid, recombined with human DNA. The desired bacteria are those in which the recombinant plasmid carrying human DNA are present which needs to be isolated for further processing.

The isolation of the desired bacterial colony can be achieved by antibiotic resistance approach. When the bacteria are incubated in a medium containing the antibiotics, ampicillin and tetracycline, all the wild bacteria will die as they are devoid of the plasmid which confer resistance to antibiotics. The surviving colonies are having either recombinant plasmid or plasmid with no human DNA. These colonies are replica plated in a medium containing chloramphenicol. Now, the bacterial colony with recombinant plasmid will die, because the human DNA is incorporated into the plasmid which carries the gene for resistance to chloramphenicol, as the gene is split by the restriction endonuclese employed and it is at this site the incorporation of human DNA occurs. Hence, the split gene, expectedly, fails to confer the resistance to chloramphenicol. The surviving colonies are those with plasmid not having human DNA, and hence, an intact gene for conferring resistance to chloramphenicol. Therefore, the desired colony are those in the original plate, corresponding to the dead colony in the replica plate. This colony can be retrieved for further processing like the processing to get the DNA, or the bacteria can be cultured to get the inserted human DNA or the corresponding protein product. In this manner the bacteria can be made to synthesise genetically engineered human proteins; for example, insulin for the treatment of diabetes mellitus without the semblance of immune response, and hence, ideally suited for human use.

The applications of DNA recombinant technology are as follows:

(a) Preparation of important biomolecules for human use, like insulin.

(b) Preparation of vaccines, like, for hepatitis B.

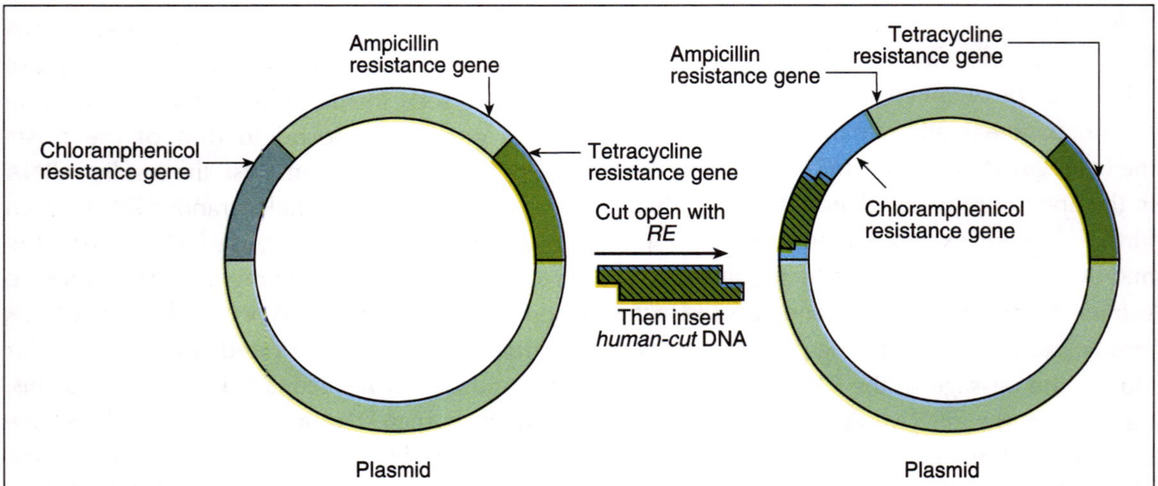

Fig. 21.2: Selection of bacterial colony by antibiotic resistance

(c) Detection of genetic disorders like HbS, cystic fibrosis, Duchene muscular dystrophy and several others.

(d) Activation of cancer genes.

(e) Mutation in the onco-suppressor genes.

(f) Presence or absence of cancer cells after treatment.

(g) Identification of criminals by analysing the DNA from blood stain, semen or hair, obtained from the crime scene and comparing the sequence with those of suspected individuals.

Animal cloning

It is scientifically possible to make an identical copy of an animal, although it is unethical to do so. The first animal cloning was done when dolly, the sheep was cloned from her mother. The dolly went on to reproduce her calf.

First a few cells from the udder of the sheep were isolated. Next the nucleus was removed from the ovum of the sheep. The denucleated ovum was then incubated with the cell obtained from the udder, when the nucleus of the cell passes into the ovum which now contains the complete DNA to produce an identical (clone) sheep. This ovum, implanted in the uterus of the sheep, grows into an identical sheep and is delivered culminating in the cloning of dolly.

A clone is an identical animal, a group identical cells or a group of identical DNA molecules, derived from a single ancestor animal, cell or DNA molecule respectively. It is to be noted that the clone cannot be fully identical because, although, the genetic information of both are the same, but for the higher animals, in general, and for humans, in particular, both intrauterine and extrauterine environment have a major say in the development of the subject and hence, the variation between an animal and its clone.

chapter 22

Polymerase Chain Reaction, Blot Techniques and Gene Therapy

POLYMERASE CHAIN REACTION (PCR)

This technique is meant for the amplification of DNA, preparing a multitude of copies of the desired DNA molecule in a very short span of time, like a photocopying machine. This is a much faster technique for gene amplification than the DNA recombinant technology. For this, we should know the sequence in the flanking regions of the target DNA to which the primer RNA has to be combined. First, the DNA strands are separated by heating at 95 degree centigrade after which it is cooled to 50 degree centigrade, then the primer, the enzyme thermus aquaticus (Taq) polymerase and the substrates for DNA synthesis are added. The temperature of the reaction system are increased to 70 degree centigrade, when the Taq polymerase polymerises deoxyribonucleotides from 5′ to 3′ direction of the growing strand as per the base pairing rules. The primer is bound to the 3′ end of the template strand of DNA and hence, the template strand of the DNA is read from 3′ to 5′.

After one round of strand separation and polymerisation we get two identical copies of the double-stranded DNA moles which will further repeat the events to form 4 double-stranded. These steps get repeated several times to make thousands of copies of the desired DNA molecule which can be used for further experiments and can be subjected to identification by blot techniques.

The Taq polymerase is a thermostable enzyme and is present in bacteria inhabiting hot water

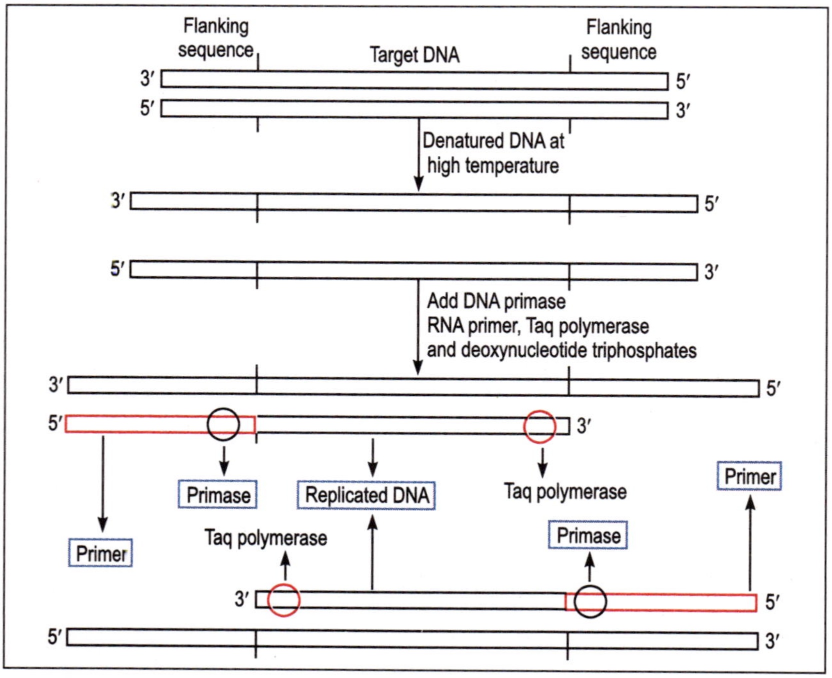

Fig. 22.1: Polymerase chain reaction

springs. Therefore, the enzyme can withstand higher temperatures and will remain active at 95 degree centigrade. The PCR can be done from mRNA, in which case the mRNA has to be first copied to complimentary DNA with another DNA polymerase enzyme isolated from a bacterium, thermus thermophilus (Tth). This is known as real time PCR (RTPCR).

The applications of PCR include:

(a) Amplification of small quantities of DNA for further research.

(b) Activation of cancer genes.

(c) Mutation in once suppressor gene.

(d) To find out whether any residual cancer cells are present after cancer therapy.

(e) Screening for hereditary diseases like cystic fibrosis, sickle cell anaemia, etc.

(f) To find out the criminals or to solve debate over parenthood.

(g) Studying the single nucleotide polymorphisms in disease processes.

Blot techniques

There are three types of blot techniques, namely Southern blot, Northern blot and Western blot, which detect specific DNA, RNA and proteins respectively.

The Southern blot technique uses single-stranded DNA probes having base complimentarily with target DNA sequences. To start the DNA molecules are subjected to digestion by restriction

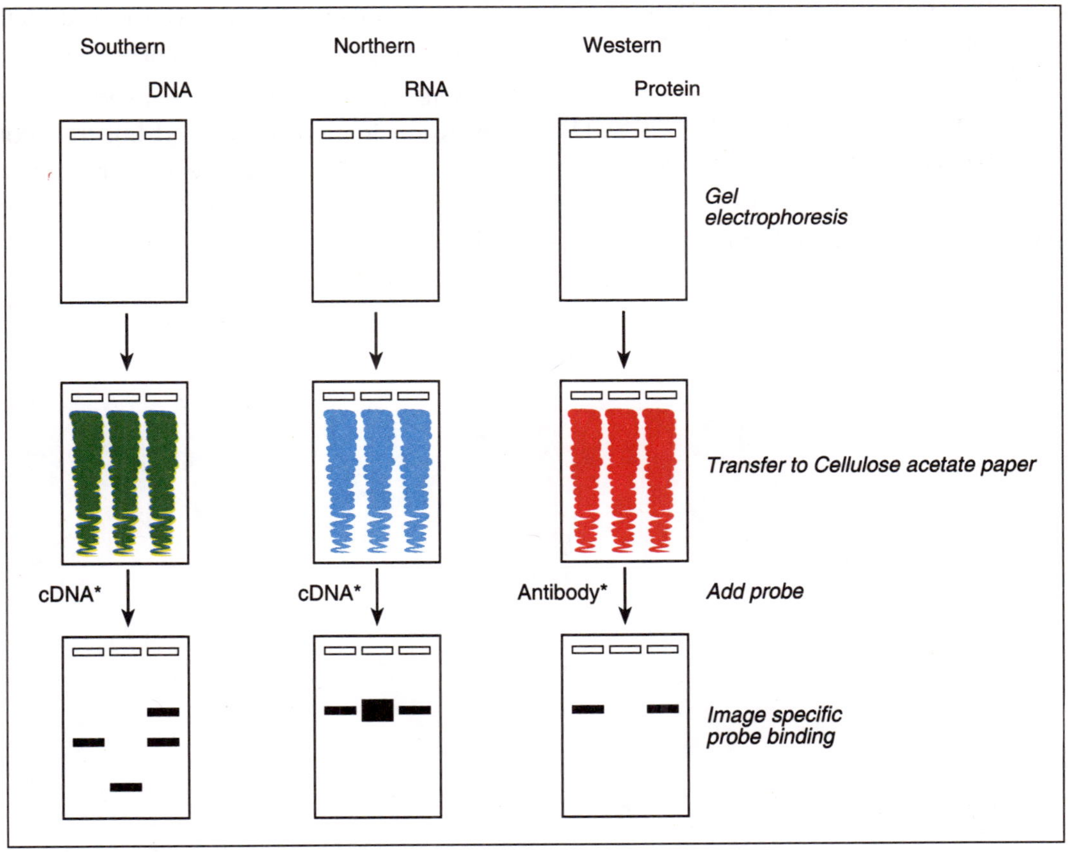

Fig. 22.2: Blot techniques

endonucleases to produce several fragments. Next these are electophoresed on agarose gel when they get separated, the smaller fragments migrate faster than the larger ones. One of these fragments could contain the target sequence (for example, a viral or a bacterial DNA or any mutated gene like Hbs, etc.). After the electrophoresis the plate is baked in an oven for fixing the fragments. Afterwards, sodium hydroxide is poured to make the DNA single-stranded. This is blotted onto a nitrocellulose paper and the radioactive probe is added. The probe is a small single stranded DNA molecule having complimentary sequence with that of the target DNA. If the target DNA is present in the mixture, the probe will bind to it. The excess of probe is washed away. The nitrocellulose membrane is placed between X-ray plates for two days when the radioactivity leaves a mark on the X-ray plate. So, Southern blot involves the detection of target DNA moles for diagnostic purposes.

The Northern blot, in principle, is same except that, here the target RNA and not DNA is detected.

In Western blot, the target protein is identified.

Gene therapy

The gene therapy is the delivery of the correct gene to a person to rectify a mutated gene. The method of delivery of the proper gene may be in vivo, ex vivo or in situ. The vectors used may be adeno virus (DNA virus) or retro virus (RNA virus). Three genes—pol, gag and env genes, should be deleted to prevent infection. The proper gene is combined with the viral DNA, and then administered to the subject. In ex-vivo method, the cells from the defective tissue are collected and the viral DNA combined with the proper gene is incorporated into the patient's cells and then put back into the patient so that the correct gene attaches with the patient's nucleus and starts the synthesis of proteins which was originally deficient in the patient.

The in vivo method is delivery of the correct gene by intravenous route. The in situ method delivers the gene directly into the tissues.

When adeno-viruses are used, the gene expression is more prolonged but the protein produced is less, whereas the retrovirus mediated delivery is transient, but proteins are formed in higher quantities.

chapter 23

Acid-base Balance

The pH of blood ranges between 7.35-7.45. It is essential that the pH of the blood has to be maintained within this narrow range, so as to make it possible for the various functions of the body to proceed at an optimum level. Life is impossible at a pH below 7 and above 7.7. Above a pH 7.5, alkalosis prevails and below 7 acidosis sets in.

During various metabolic steps certain acids are generated which need to be neutralised, lest the pH remains within the normal range. The H_2CO_3 is the major acid produced during cellular metabolism. Other acids like hydrochloric acid, sulphuric acid, phosphoric acid, lactic acid, pyruvic acid and uric acid are also generated. Some quantities of bases are also formed in the reactions.

As these acids and bases get entry into the blood, the body has an important job of maintaining the pH within the normal range, with minimum possible fluctuations. The body achieves this by the following mechanisms:

1. Blood buffer system
2. Respiratory system and
3. Excretion of acids or bases by the kidneys, as the case may be, and further by the synthesis and excretion of ammonia.

Blood buffers

The blood buffers are the first line of defence against alternations in pH. The buffers are mixtures of weak acids and their salts.

A base can be defined as one that combines with hydrogen ion and an acid as one that releases hydrogen ions. If we look at the reaction of dissociation of H_2CO_3, we can write as:

$$H_2CO_3 \rightarrow H^+ + HCO_3^-$$

Out of this, H_2CO_3 which releases H^+ is the acid and HCO_3^- which binds to H^+ is the base.

The lungs and the kidneys are the organs to excrete acids and bases. The major blood buffers are bicarbonate, phosphate, protein and haemoglobin of which the haemoglobin buffer is in the RBC. The various buffers are written as:

1. The bicarbonate buffer: $NaHCO_3/H_2CO_3$ of which $NaHCO_3$ is the salt and H_2CO_3 is the acid.
2. The phosphate buffer: Na_2HPO_4/NaH_2PO_4.
3. The protein buffer: Na-Protein/H-Protein.
4. The haemoglobin buffer: KHb/HHb.

The bicarbonate and protein buffers are present in the plasma. The RBCs also contain bicarbonate buffer and, additionally, haemoglobin buffer. The phosphate buffer concentration being lower in plasma, is not that effective in plasma. The bicarbonate buffer system, present in both erythrocytes and plasma, is most effective in buffering all types of acids, stronger than carbonic acid.

If there is ketoacidosis due to starvation or diabetes mellitus, the strong acetoacetic acid, produced in large amounts, as a result of increased lipolysis and beta-oxidation of fatty acids, is neutralised with the help of $NaHCO_3$ of bicarbonate buffer and forms sodium acetoacetate and carbonic acid. By this way, the strong acetoacetic acid is transformed into the weak carbonic acid, resulting in hardly any change in pH. All the other acids also could be neutralised in this manner.

Similarly, an influx of an alkali like NaOH, can be countered by H_2CO_3, which can react with the alkali and form $NaHCO_3$ and H_2O.

The CO_2 present in the blood combines with water to form carbonic acid which dissociates to form HCO_3 and H^+.

The ratio between sodium bicarbonate to carbonic acid is 20:1. The bicarbonate is the most important buffer and does about 60 per cent of the buffering in the plasma. Other buffers include phosphate and protein buffers.

Respiratory mechanism

The respiratory centre in the brain has chemoreceptors which are sensitive to alterations in pCO_2 and pH. When acidosis occurs due to an increase in carbon dioxide, the respiratory centre is activated and causes deep breathing by hyperventilation. This will increase the CO_2 elimination resulting in lowering of carbonic acid and the acidosis is correct.

A rise in the CO_2 in the tissues, releases the oxygen from haemoglobin. The CO_2 diffusing out of cells combines, in presence of carbonic anhydrase, with water to form carbonic acid which dissociates into H^+ and HCO_3^-. The haemoglobin combines with H^+ to form HHb.

In the lungs, the reverse takes place, namely, the oxygen binds with haemoglobin when H^+ is released. The H^+ combines with HCO_3^- to form H_2CO_3 that breaks down to H_2O and CO_2. The carbon dioxide is removed by expiration.

Renal regulation of acid-base balance

The pH of urine can vary from acidic to alkaline depending on the acidification of urine. The pH of blood is around 7.4 but the pH of urine is usually about 6. The kidneys serve to regulate pH as follows:

1. By excreting hydrogen ions, formed by the dissociation of carbonic acid, into the proximal convoluted tubular lumen in exchange of sodium which is reabsorbed along with bicarbonate. The carbonic acid is generated by carbonic anhydrase in presence of water and CO_2. The secretion of hydrogen ions along with the reabsorption of $NaHCO_3$ ultimately restores the base reserve.
2. There is net excretion of acids in the distal convoluted tubules.
3. The excretion of ammonium ions also takes place in the distal convoluted tubules. The glutaminase enzyme acts on glutamic to form ammonia and glutamic acid. The ammonia produced diffuses into luminal fluid where it combines with hydrogen ions to form ammonium ions which can replace the sodium and potassium from salts, thereby preserving alkali reserve.

Acid-base balance disorders

In an acidotic state, either acids accumulate or the bases are lost. Alkalosis prevails when the reverse occurs, that is, acids being lost and bases being retained. The maximum pH alteration that the body can tolerate is between 7-7.6.

Acidosis

The acidosis can be of two types: respiratory and metabolic. The cardinal feature of the respiratory acidosis is an excess of carbonic acid (pCO_2 > 45mm Hg) and that of the metabolic acidosis is bicarbonate deficit (HCO_3 < 22mmol/L). The acidosis is achieved when the pH is lowered.

Alkalosis

The alkalosis can also be two types: respiratory

alkalosis due to the deficit of carbonic acid (pCO_2 < 35mm Hg) and metabolic alkalosis due to the excess of bicarbonate (HCO_3 > 33mmol/L).

Compensatory mechanisms

The body tries to counter acidosis and alkalosis by compensatory mechanisms. The general basis of compensation can be stated as follows: If we consider that an abnormal codition has changed one of the components of HCO_3^-/H_2CO_3, say the bicarbonate, the pH can be returned to normal by changing the H_2CO_3 (pCO_2) also in the direction.

When the HCO_3 levels are decreased as in metabolic acidosis, the body tries to lower the pCO_2 levels also as a compensatory mechanism.

In respiratory acidosis, there is high pCO_2 content, which the body tries to compensate by increasing the HCO_3 level also.

In metabolic alkalosis, there is increased HCO_3 level which the body tries to compensate by increasing the pCO_2 levels also by hypoventilation.

In respiratory alkalosis there is decrease of pCO_2 content which the body tries to compensate by decreasing the level of HCO_3 also.

Due to the above cited mechanisms the acidosis or alkalosis may be uncompensated, partially compensated or fully compensated.

Normal values

The normal levels of arterial blood gases and pH are:

(a) pH: 7.34-7.45.

(b) pCO_2: 35-45mm Hg

(c) pO_2: 80-100mm Hg

(d) HCO_3^-: 22-26meq/L

Generally stating, acidosis is accompanied by hyperkalaemia and alkalosis is associated with hypokalaemia.

Anion gap

To maintain electrical neutrality, the sum of anions and cations are equal in extracellular fluid. Sodium and potassium are the major cations and together constitute around 90 per cent of the cations meanwhile bicarbonate and chloride contribute about 80 per cent of the anions. Usually, only these electrolytes are estimated. This means there is a difference (gap) between the measured cations and anions. This gap is filled by the proteins and other anions which represent the unmeasured anions. This is known as the anion gap which normally is about 12mmol/L. The gap may be increased when the cations are decreased.

chapter 24

Thyroid Hormones and Thyroid Function Tests

The thyroid gland can take up iodine. The iodine content of the body is about is 35mg. It can be obtained from water cereals and vegetables. The high altitude regions have less iodine. Thiocyanate and thiourea adversely affect iodine metabolism.

Synthesis of thyroid hormones

The synthesis of thyroid hormones starts with the uptake iodine by the thyroid gland. This process of uptake of iodine is facilitated by the thyroid stimulating hormone (TSH), secreted by the anterior pituitary gland.

In the next step, requiring NADPH, the thyoperoxidase enzyme oxidises iodine which is also under the stimulatory effect of TSH.

The third step is that of iodination of the protein thyroglobulin, formed in the thyroid gland. The thyroglobulin, one of the largest of all proteins, has several tyrosine residues of which about 30-40 tyrosine residues are iodinated. This can result in mono-iodotyrosine, and di-iodotyrosine are synthesised.

The fourth reaction is the coupling of 2 di-iodotyrosine molecules which remain juxtaposed to generate a tetra-iodothyroine (T_4). The tri-iodothyronine (T_3) is synthesised by deiodination of T_4. The thyroid gland can store T_4 and T_3 as a colloid.

When there is a need for the hormone, the protease enzyme releases it from the thyroglobulin, a process that TSH stimulates. Abnormality occurs when the T_4 cannot be liberated from a defective thyroglobulin. The T_3 is produced from T_4 by the removal of one iodine atom. At times, the process of deiodination can also be defective.

The T_3 and T_4 are transported by binding to specific plasma proteins, the thyroid binding globulins. The plasma albumin and transthyretin also help in the thyroid hormone transport. The thyroid hormones attach to the hormone response elements and bring forth its metabolic effects which include protein catabolism and heat generation. It increases basal metabolic rate and cellular metabolism. The cholesterol level is decreased.

Assessment of thyroid function

This is done by assaying the thyroid hormone levels in the serum. Hence, T_3 and T_4 levels, along with the anterior pituitary hormone, TSH are estimated.

In hyperthyroidism, serum T_3 and T_4 levels are elevated, but TSH is decreased. The hypothyroid patients have lower levels of T_3 and T_4 and raised concentrations of TSH. This happens in primary hypothyroidism, but in secondary hypothyroidism due to pituitary and hypothalamic defect, in addition to a decrease in T_3 and T_4. The TSH is also lowered. As the levels of T_3 and T_4 can be influenced by the status of carrier proteins, new method of assay measures the free T_3 and free T_4 forms which represent the active hormones, and hence, are better indicators of thyroid function.

The cholesterol level is found to be lowered in hyperthyrodism and elevated in hypothyroidism. The basal metabolic rate is decreased in hypothyroidism and increased in hyperthyroidism.

The examples of hyperthyroidism are Grave's disease and T_3 toxicosis.

An autoimmune condition affecting thyroid can lead to primary hypothyroidism. The hypothyroidism is more common in the females.

chapter 25

Kidney Function Tests

The major functions of kidneys are regulating osmotic pressure, water, electrolyte and acid-base balance of the body, in addition to eliminating the waste and toxic substances from the body. Many substances present in urine are in higher concentration as compared to blood, say urea, creatinine and uric acid, while certain others are absent (glucose) or in trace amounts.

Formation of urine

Two major steps are involved in the urine formation, namely glomerular filtration and tubular reabsorption and accompanying concentration of urine.

Glomerular filtration (GFR)

The urine is the ultrafiltrate of blood and so the step in urine formation is the filtration of nonprotein substances across the glomerular membrane, that means it prepares a protein-free filtrate of the waste products arising out of metabolic reactions. The composition of glomerular filtrate resembles that of plasma minus the proteins. The effective filtration pressure at the glomerulus is contributed by the blood pressure in the glomerular capillaries which has to be greater than the total of the oncolytic pressure and the pressure in the Bowman's capsule. The glomerular filtrate formed per minute is about 120 to 130ml and the volume of glomerular filtrate generated is about 165 litres everyday. The factors affecting the renal blood flow can affect the filtration rate, like conditions of cardiac failure, low blood pressure and dehydration.

Tubular reabsorption

The tubular reabsorption occurs when the glomerular filtrate passes through the small uriniferous tubules in the kidneys. In these tubules a major portion (99 per cent) of water is reabsorbed along with glucose, electrolytes and phosphates. The proximal tubules reabsorb around 75-80 per cent and the distal tubules about 15-20 per cent of the glomerular filtrate. This is accompanied by the concentration of urine. This step is regulated by antidiuretic hormone, secreted by the posterior pituitary gland. Normally, about 1.5 litres of urine with a specific gravity of 1.01-1.03 per day are eliminated from the body.

Glucose reabsorption

The arterial blood has a glucose level of about 120mg/dl, which is filtered by the glomerulus, but this whole amount is reabsorbed in the proximal tubules. The tubular maximum for glucose (TMG) absorption is 350mg per minute. If glucose, filtered, is more than capacity of the tubules to reabsorb (350mg/minute), the excess amount appears in the urine (glycosuria). The glucose excretion entails associated elimination of water also resulting in polyurea which is a salient feature of diabetes mellitus.

Electrolytes

Sodium and potassium are filtered at the glomerulus and as much as 99 per cent is reabsorbed in the kidney tubules, the process being regulated by aldosterone. Most of the sodium is reabsorbed along with chloride, with some sodium absorbed in exchange of H^+ ions for the acid base balance. The secretion of H^+ in exchange of sodium in the distal tubules is a step required for the acidification of urine.

Potassium is also extensively subjected to maximal reabsorption from the kidney tubules, but some potassium is secreted by distal tubules also.

The normal urine is almost devoid of bicarbonate, as it is reabsorbed. Similarly, amino acid excretion in urine may be in traces.

Urea, uric acid and creatinine

Waste materials, urea and uric acid are also reabsorbed to different degrees. The creatinine, another metabolic waste, is excreted from glomerulus and also from proximal tubules.

Inulin

The inulin excretion is taken as an index of GRF, because it is fully filtered at the glomerulus but not all reabsorbed in the tubules.

Renal function tests

The renal function tests include estimation of urea and creatinine in the plasma. The urea is estimated by urease method or diacetylmonoxime method. The creatinine is estimated by Jaffe kinetic method or enzymatic method. In addition there are the clearance tests like urea clearance, creatinine clearance and inulin clearance to assess the GFR. The inulin clearance test is the preferred one because inulin is not only filtered fully at the glomerulus, but is not reabsorbed at all in the tubules. GFR value less than 60mL/min/173m square indicates chronic kidney disease.

Normal levels of serum urea and creatinine

The urea and creatinine tests are very important in gauging the severity of kidney related diseases. Their normal values are as under:

Serum urea: 20-40mg/dl.

Serum creatinine: Males: 0.7-1.4mg/dl

Females: 0.6-1.3mg/dl.

The values above this are indicative of kidney dysfunction. Hence, it is necessary that the diabetic patients should undergo the test of serum urea and creatinine every 6 months as long term hyperglycaemia can cause kidney damage.

Cystatin C

The cystatin C is a new marker of GFR. It is a small polypeptide and is a cysteine proteinase inhibitor. The cystatin C is found to increase even before creatinine shows an increase. The increase of creatinine occurs only when the GFR is moderately decreased, while the cystatin C is elevated with very slight decrease of GFR.

Proteinuria and microalbuminuria

Normal urine does not contain proteins. As the glomerulus gets damaged, albumin with a molecular weight of 69,000 filtered at the glomerulus and causes albuminuria. The proteins are excreted in the urine by patients with nephritis.

Microalbuminuria is a condition when small quantities (30-300mg/day) of albumin are excreted in the urine. This is a very important test to be carried out in diabetes mellitus patients so as to know about the status of their kidneys. If microalbuminuria is detected in a patient, he should try to control the blood glucose levels stringently.

chapter 26

Biochemistry of Cancer and Cancer Markers

Cancer is caused by a variety of factors including genetic, (presence of oncogenes, absence of oncosuppressor genes), environmental, (mutagens, pollutants), lifestyle (smoking, high fat food, alcohol, sedentary life, obesity, etc.) and metabolic. Aging increases the occurrence of cancer as the mutation rate is increased in old age and some of these mutations can be cancer causing.

Agents that cause cancer

Carcinogens induce mutations which initiates the transformation of a normal cell into a cancer cell. The single cancer cell multiplies to create a cancer mass. Some of the mutagens can cause cancer. Components of tobacco and asbestos are carcinogenic, so are UV rays and ionic radiations like X-rays and gamma rays. Several hundreds of chemicals are cancer causing. The oncogenic viruses also can start carcinogenic processes.

1. The aflatoxins present in certain moulds which grow on cereals, if consumed, may lead to cancer.
2. The components of cigarette like carbon soot, benzopyrene and nicotine amongst many others, can result in cancer. Cigarette smoking considerably reduces one's lifespan as it causes not only cancer but also coronary heart disease. The people who are present in the vicinity of a smoker are equally at risk of developing cancer, and hence, the close relatives and friends also could be afflicted with this dreaded ailment. The cigarette smoking leads to cancer of the lungs, whereas tobacco chewing is responsible for oral cancer.
3. Excessive intake of alcohol is implicated in the development of cancers of mouth, oesophagus and liver as it causes liver cirrhosis.
4. A high fat diet, containing very large quantities of cholesterol is another carcinogenic factor, as it may cause colon cancer.
5. A diet containing less fibres, leafy vegetables and fruits (the latter two having antioxidants) increases the cancer risk. Vitamins like vitamin A, betacarotene, ascorbic acid and vitamin E help to scavenge the free radicals, thereby protecting the body from the bad effects of free radicals. The turmeric powder prevents mutations so also the flavonoids seen in plant foods. Both have anticancer properties. The different types of berries are rich in phenolic compounds which also have antimutagenic properties.
6. A reduction in the antioxidant enzymes, like glutathione peroxidase, glutathione reductase and glutathione-s-transferase could also lead to cancer development as they are involved in the scavenging of free radicals.
7. The oncogenic viruses may get incorporated into the human genome and this can cause oncogenesis. Some examples of oncogenic viruses are hepatitis B virus, human papilloma virus, etc.
8. The precursors of carcinogens taken inadvertently in the foods get metabolised to the active forms in the liver, and are responsible for carcinogenesis. The electron-deficient carcinogens react with electron-rich purines, causing damage to the DNA. The cells try to repair the damage to DNA, which when

unsuccessful, results in mutations. These mutations subsequently lead to a malignant state.

9. The ionising radiations like X-rays and gamma rays cause damage to the DNA, which when not repaired, can cause malignancy. The UV light can initiate the formation of purine-purine dimers which is the cause of one type of skin cancer.

Oncogenes

The oncogenes are cancer producing genes. The oncogenes may remain as proto-oncogenes which get activated to oncogenes. Factors like oncoviruses, ionising radiations and chemical carcinogens activate oncogenes which cause malignancy. It may be activated by mutations in antioncogens. The ras gene is an oncogene.

Antioncogenes or oncosuppressor genes

These genes protect the people from getting cancer. The p53 is an oncosuppressor gene whose gene product is p53 protein. The p53 protein prevents the division of cells containing damaged DNA till the damaged DNA is fully repaired. If the damaged DNA is irreparable, p53 instructs the cells to proceed for apoptosis or programmed cell death. Cancerous cells do not have p53, so that these cells cannot repair DNA damage or undergo apoptosis. Hence, cancer cells are immortalised.

Another oncosuppressor gene is retinoblastoma gene (RB gene) which also inhibits cell division. Mutations in the RB and p53 genes lead to cancer production.

The cancer cell membrane contains higher quantities of N-acetyl neuraminic acid due to which cancer cells have a negative surface. This property make the cancer cells repel which can explain metastasis of cancer. The collagenase present in the cancer tissue degrades the connective tissue collagen which further facilitates the process of metastasis.

The telomeres and telomerase in cancer cells

Every cell on division loses some of its DNA from the unreplicated 3′ end of the parent DNA strand which had bound to the primer RNA while initiating the replication process. This region of the parent strand, known as the telomere, remains single stranded after the removal of the primer RNA. The telomere can be restored in length with the help of telomerase. If the telomerase activity is decreased, as in old age, the length of the chromosomes gets shortened. This is in contrast to cancer cells where the telomerase activity is high.

Tumour markers

The tumour markers are substances, produced by the cancer tissue, the presence of which can suggest a carcinogenic state.

The cancer marker substances cannot be used for diagnosing cancer because certain nonmalignant conditions also may lead to the appearance of tumour markers in the serum. Further, some of the cancers produce tumour markers only inconsistently, thereby limiting their diagnostic value.

Tumour markers provide supportive evidence to biopsy and imaging studies. The tumour markers are of utility in monitoring whether the treatment modalities are showing improvement in the patient or not. It should also be kept in mind that most of the cancer markers are nonspecific showing increase in a variety of cancers while trying to interpret the results.

Some of the useful tumour markers are:

1. Alpha fetoprotein is found to be elevated in ovarian and hepatocellular cancers.
2. Carcino-embrionic antigen (CEA) is found to be elevated in different types of cancer like ovarian, lung, pancreatic, colorectal cancers, etc.
3. Cancer antigen 125 is marker for ovarian cancer. Elevated levels may be encountered in certain benign gynaecological states also, say ovarian cysts.
4. Human chorionic gonadotropin (HCG) serves to maintain the corpus luteum during pregnancy. Elevated levels of HCG during pregnancy are of value in diagnosing chorionic carcinoma.
5. The prostate specific antigen (PSA) is elevated in the cancer of prostate glands.
6. Beta macroglobulins are found to be elevated in multiple myeloma.
7. The calcitonin which is a hormone of the thyroid gland, helping in calcium homeostasis, serves as a marker of thyroid carcinoma.
8. Thyroglobulin also is reported to increase in thyroid carcinoma.

Drugs used for cancer treatment

There are several drugs helpful in the treatment of cancer although the most effective line of treatment is still surgery and radiotherapy. The chemotherapeutic drugs are mostly analogues of nucleosides, which, after getting incorporated into the DNA, stop the process of replication of DNA, resulting in arrest of cell proliferation. Alternatively, the nucleoside analogues may get metabolised to compounds which inhibit enzymes involved in various cell functions.

Some examples of anticancer drugs include:

1. 6-mercaptopurine that prevent the formation of AMP.
2. 5-fluorouracil and iodouracil, serving as analogues of thymine.
3. Thioguanine, a purine analogue.
4. Methotrexate, an inhibitor of dihydrofolate reductase required for the formation of active THFA that donates the C_2 and C_8 of purines.
5. Azaserine is an analogue of glutamine which provide the N_3 and N_9 of purines.
6. Asparginase is an enzyme degrading the amino acid asparagine, thereby reducing the supply of this amino acid to cancer cells. Asparagine can be formed by normal cells as it is an essential amino acid.
7. Cytarabine and viderabine which have arabinose as the pentose in place of ribose.

Certain plant alkaloids and some other alkylating agents forming cross-linkages between bases and between the two strands of DNA also find their use as anticancer agents. A recent addition to these drugs are monoclonal antibodies.

chapter 27

Mechanisms of Action of Hormones

The endocrine system consists of a group of glands that produces regulatory chemicals called **hormones**. The endocrine and the nervous systems function in tandem to regulate all the other systems of the body.

Hormones are biochemical messengers that have specific regulatory effects on certain cells and organs. They are secreted from the endocrine glands and released directly into the blood stream, which carries them all parts of the body. Their functions include activating the processes of transcription and translation.

Based on the mechanisms of action, the hormones may be classified into two groups:

1. Hormones acting through intracellular receptor.
2. Hormones acting through cell surface receptor.

Hormones acting through intracellular receptor

The hormones in this group include the steroid hormone and thyroid hormone.

Process

1. Hormone diffuses through the membrane.
2. It binds and activates intracellular receptor.
3. Steroid-receptor complex binds to DNA receptor protein.
4. Activates a gene.
5. Gene transcribed into messenger RNA.
6. mRNA goes to the ribosomes.
7. Translates mRNA into protein.

Hormones acting through cell surface receptor

There are three types of cell surface receptor

1. Ion channel receptors
2. Transmembrane receptors
3. Receptors that are kinases or bind with kinases.

Ion channel receptors

When a hormone binds to an **ion channel** on the outside of the cell, they open the channel and allow the ions to move in or out of the cell following their electrical gradients and thus altering the polarization of the cell membrane.

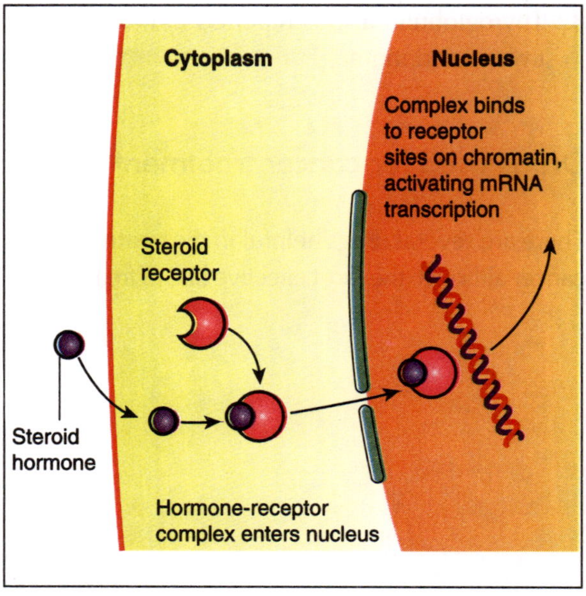

Fig. 27.1: Thyroid hormones and retinoids go directly into the nucleus

Receptors that are kinases or bind with kinases

When a signalling molecule binds to the membrane receptor protein on the outer portion of the cell, they activate kinases, which in turn trigger a chemical reaction on the inside of the cell.

Transmembrane proteins

Transmembrane proteins include G-protein-linked receptors and they are seven-pass transmembrane proteins. This means that the polypeptide chain trasverses the membrane seven times. When hormone binds to the receptor on the outer portion of the cell, this triggers a series of chemical reactions like the movement and binding of the G-protein, conversion of GTP into GDP and activation of second messenger which are molecules that relay signals from receptors on the cell surface to target molecules in the cytoplasm or nucleus.

According to second messenger and enzyme activation, the hormones are classified into five groups.

Group	Second messenger	Examples of hormone
A	cAMP	TSH, Glucagon, ACTH, LH, FSH, HCG
B	cGMP	Atrial natriuretic factor, nitric oxide
C	Calcium and phosphatidyl inositol	TRH, Acetylcholine, gastrin, CCK, Oxytocin
D	Tyrosine kinase	Insulin, growth factor
E	Kinase or phosphatase cascade	Adiponectin, leptin, GH, TNF, erythroprotein

Hormones Acting Through G-Protein

The G-complex is composed of three subunits α, β and γ in which the α-subunit is either stimulatory or inhibitory and this binds to GDP or GTP.

When the hormone binds to the receptor, certain conformational change occurs in the G-

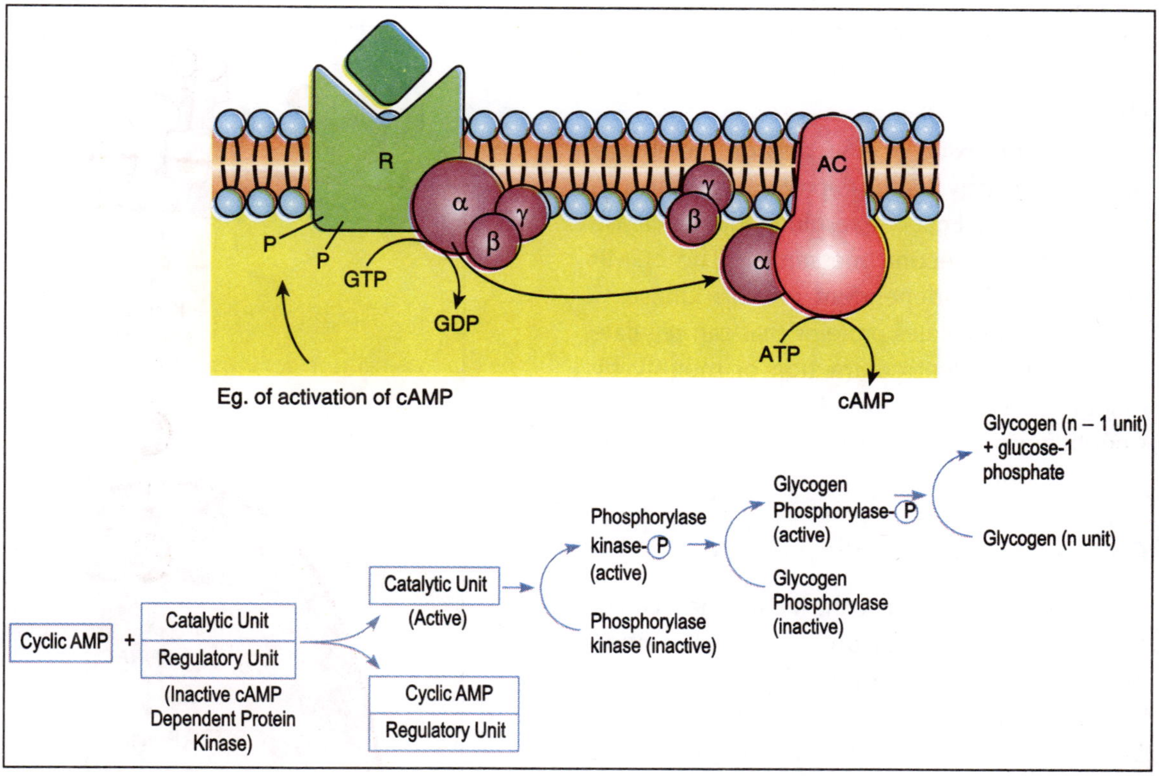

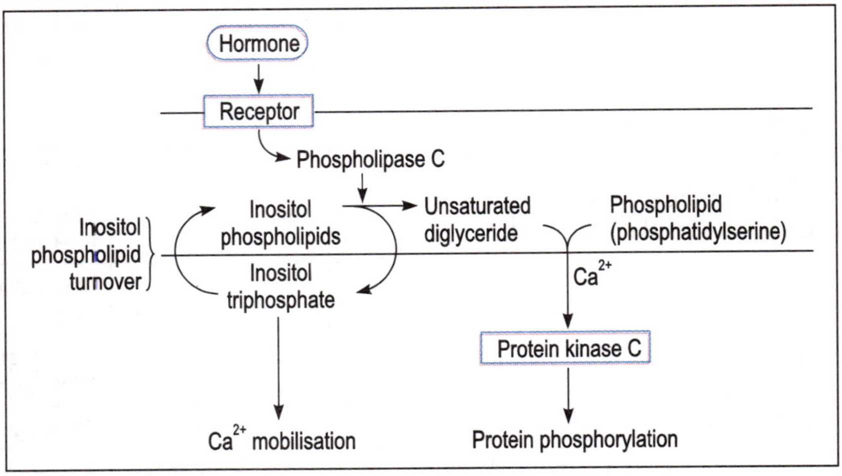

complex and the alpha subunit binds GTP instead of GDP and the alpha subunit dissociates from β and γ subunits. The αs (α stimulatory subunit) protein has intrinsic GTPase activity and it catalyses the conversion of GTP to GDP.

The activated G-protein binds to the target protein which results in its altered activity resulting in either stimulation or inhibition.

The different peptide hormones can either stimulate or inhibit the production of cAMP from adenylyl cyclase (AC).

The nitric oxide or ANF acts on membrane receptor and activates the G-protein which in turn activates Guanyl cyclase resulting in an increase of cGMP as a second messenger. The cGMP activates cGMP-dependent protein kinase G which phosphorylates proteins that can regulate calcium dependent contraction or motility by modulating calcium influx. An example is that of smooth muscle myosin, leading to relaxation and vasodilatation.

Calcium and phosphatidyl inositol

The binding of serotonin to the membrane receptor triggers the activation of the enzyme phospholipase C which hydrolyses the phosphatidyl inositol to diacylglycerol and 1, 4, 5-inositol triphosphate (IP_3), which can release calcium from intracellular store and this triggers the processes like smooth muscle contraction, glycogen breakdown and exocytosis.

Tyrosine kinase activation

- Insulin binds on alpha subunit membrane receptor and phosphorylate beta subunit having tyrosine kinase activity.

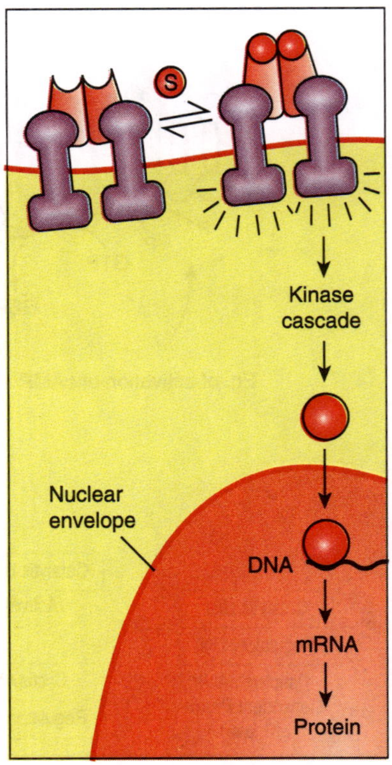

- Activated tyrosine kinase initiates a kinase cascade.
- IRS (insulin response substrate) is phosphorylated.
- At gene level.
- Induces glucokinase, pyruvate kinase, phosphofructokinase and acetyl CoA carboxylase.
- Represses glucose 6-phosphate, phosphoenol pyruvate carboxykinase, fructose 1, 6-bisphosphatase.
- Insulin activates some enzyme by covalent modification.
- Insulin increases the recruitment of GLUT 4 in cell and increases glucose uptake.

chapter 28

FREE RADICALS AND ANTIOXIDANTS

A free radical is defined as any molecular species capable of independent existence that contains an unpaired electron in an atomic orbital. Free radicals are unstable and highly reactive. They can either donate an electron to or accept an electron from other molecules, thereby behaving as oxidants or reductants. Free radical is generally represented by a superscript dot ($R^\bullet$).

Reactive oxygen species (ROS) are oxygen containing molecules, which are more active than the oxygen molecule present in air.

ROS include free radicals like:

- Superoxide (O_2^-)
- Hydroxyl radicals ($OH^\bullet$)
- Peroxyl radical ($RO_2^\bullet$)
- Hydroperoxyl radical ($HOO^\bullet$)
- Singlet oxygen ($^1O_2$)

Reactive nitrogen species (RNS) include free radicals like:

- Nitric oxide ($NO^\bullet$)
- Nitrogen dioxide ($NO_2^{-\bullet}$)
- Peroxynitrite ($ONOO^{-\bullet}$)

The non-radical species include:

- Hydrogen peroxide (H_2O_2)
- Hypochlorous acid (HOCl)
- Nitrous oxide (HNO_2)
- Alkyl peroxynitrates (RONOO)

GENERATION OF FREE RADICALS

Free radicals are continuously produced in the body.

1. They can be formed as a by-product during the normal oxidation of foodstuffs in the mitochondrial respiratory chain due to leakage of electrons from the electron transport chain or by reactions catalyzed by transition metal ions such as iron and copper ions.
2. They may also be derived from external sources such as cigarette smoke, radiation, UV light, pollution and from the metabolism of certain drugs.
3. Free radicals in addition to possessing deleterious effects, do help in immune functions. A well-defined role for free radicals is when activated phagocytic cells

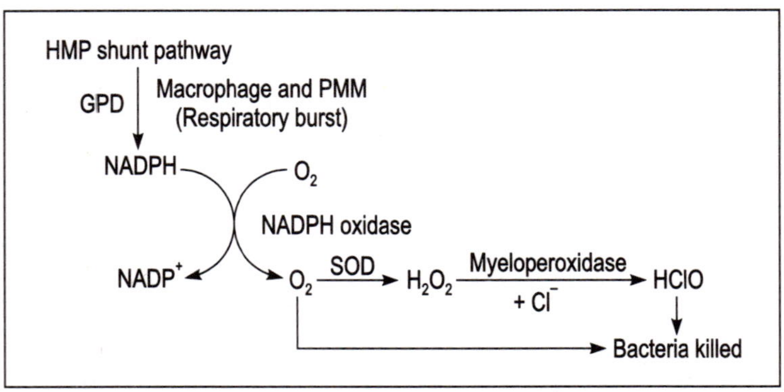

(neutrophils, monocytes, macrophages and eosinophils) produce superoxide anion radicals and hydrogen peroxide as one mechanism to kill bacteria and fungi and to inactive viruses. Along with the activation of macrophages, there is a drastic increase in the oxygen consumption by the cell. This is called as respiratory burst.

FREE RADICAL SCAVENGER SYSTEM

Antioxidants

Antioxidants are defined as compounds that inhibit or delay the oxidation of other molecules by inhibiting the initiation or propagation of oxidizing chain reactions. They are also called as oxidation inhibitors. At any point of time, one antioxidant molecule can react with single free radical and is able to neutralize free radical(s) by donating one of their own electrons. Antioxidants prevent cell and tissue damage as they act as scavenger. A variety of components act against free radicals to neutralize them from both endogenous and exogenous origin.

These antioxidant systems are classified as below:

(a) Enzymatic antioxidant systems

These are preventive antioxidants which inhibit the production of free radicals.

1. **Superoxide dismutase (SOD)**: Mitochondrial SOD is manganese dependent whereas cytoplasmic SOD is copper-zinc dependent. The reactions catalysed by these enzymes are as following:

$$2O_2^{\bullet} + 2H^+ \xrightarrow{\text{SOD}} H_2O_2 + O_2$$

2. **Catalase**: When H_2O_2 is produced in large amounts, the enzyme catalase is also used for its removal.

$$2H_2O_2 \xrightarrow{\text{Catalase}} 2H_2O + O_2$$

3. **Glutathione peroxidase (GSHPx)**: In the next step, H_2O_2 is removed by glutathione peroxidase (GSHPx).

$$\underset{\text{(Reduced)}}{2GSH} + H_2O_2 \xrightarrow{\text{GSHPx}} \underset{\text{(Oxidised)}}{CSSG} + 2H_2O$$

4. **Glutathione reductase (GR)**: The oxidized glutathione is reduced by glutathione reductase (GR) which requires NADPH from the HMP shunt pathway. So in G-6PD deficiency, the RBCs are liable to lysis especially when oxidizing agents like primaquine, antibiotic and antimalarial drugs are given. This leads to drug induced haemolytic anaemia.

$$\underset{\text{(Oxidised)}}{GSSG} + NADPH + H^+ \longrightarrow \underset{\text{(Reduced)}}{2GSH} + NADP^+$$

(b) Non-enzymatic antioxidant system

They are chain breaking antioxidants which inhibit propogative phase of free radical generation.

1. Endogenously produced antioxidants like glutathione, ubiquinol, uric acid.
2. Derived from diet like vitamin E, vitamin C, carotenoids, lipoic acid, selenium, etc.

The mechanism by which non-enzymatic chain breaking antioxidants work is shown as following:

$$ROO^{\bullet} + \text{Vit. E—OH} \rightarrow ROOH + \text{Vit. E—O}^{\bullet}$$

$$\text{Vit. E—O}^{\bullet} + \text{Vit. CH} \rightarrow \text{Vit. E OH} + \text{Vit. C}^{\bullet}$$

DAMAGE PRODUCED BY FREE RADICALS

The main biological targets for attacks by free radicals are:

1. **Lipids**: Free radicals cause peroxidation of PUFA (polyunsaturated fatty acid) in the plasma membrane. This leads to loss of membrane functions.

Three phases are involved:

(a) Initiation phase

$$RH + OH^{\bullet} \rightarrow R^{\bullet} + H_2O$$

$$ROOH \rightarrow ROO^{\bullet} + H^{+}$$

(b) Propagation phase

$$R^{\bullet} + O_2 \rightarrow ROO^{\bullet}$$

$$ROO^{\bullet} + RH \rightarrow ROOH + R^{\bullet}$$

(c) Termination phase

$$ROO^{\bullet} + ROO^{\bullet} \rightarrow RO\text{–}OR + O_2$$

$$R^{\bullet} + R^{\bullet} \rightarrow R\text{–}R$$

$$ROO^{\bullet} + R^{\bullet} \rightarrow RO\text{–}OR$$

2. **Proteins**: Free radicals cause oxidation of sulfhydryl group containing enzymes, thereby causing loss of function and fragmentation of protein.
3. **DNA damage**: DNA is damaged by strand breaks. The DNA damage may directly cause inhibition of protein and enzyme synthesis, and thereby indirectly causes cell death or mutation and carcinogenesis.

DISEASES CAUSED BY FREE RADICALS

1. *Eye*: Diabetic retinopathy of prematurity, cataracts, age related macular degeneration.
2. *Brain*: Parkinson's disease, Alzheimer's disease, amyotrophic lateral sclerosis.
3. *Respiratory tract*: Bronchial asthma, inhalation injury.
4. *Chest*: Cardiac arrhythmia, cardiac infarction, high blood pressure.
5. *Abdomen*: Gastric ulcer, ischaemic colitis, fatty liver, kidney failure, uraemia.
6. *Generalized body*: Aging, diabetes, allergy, rheumatic disease, cancer, arterial sclerosis.

chapter 29

Detoxification and Biotransformation of Xenobiotics

A **xenbiotic** (Gk. *xenos* "stranger") is a compound that is foreign to the body which may be accidentally ingested or taken as drugs or compounds produced in the body by bacterial metabolism. Molecules to be eliminated / detoxified are called xenobiotics. Examples are drugs, food additives, pollutants, insecticides, chemicals carcinogens, bilirubin, bile acids, steroids, eicosanoids, etc.

Biotransformation is a process whereby a substance is changed from one chemical to another by a chemical reaction in the body. **Detoxification** is a process whereby the noxious substances are converted to less harmful and more water soluble substances. Biotransformation is not exact synonymous of detoxification. In some cases the metabolites are more toxic than the parent compounds, e.g., biotransformation of vinyl chloride to vinyl chloride epoxide that causes cancer of liver. Liver plays important role in detoxification reactions.

Metabolism of Xenobiotics

Metabolism of xenobiotics occurs in two phases:

- **Phase 1 reaction**: In this reaction functional group is added to the xenobiotic.

Phase 1 reaction include:

(A) Oxidation

(B) Reduction

(C) Hydrolysis reactions.

- They are also called **Hydroxylation reactions** since they introduce or expose a functional group (e.g., —OH) that serves as the active centre for sequential conjugation in a phase II reaction.

(A) Oxidation

A large number of foreign substances are destroyed by oxidation in the body.

Examples:

- Oxidation of methyl group containing compounds:

  Methyl group is oxidized to acid through formation of alcohol and aldehyde:

  $CH_3 \rightarrow CH_2OH \rightarrow CHO \rightarrow COOH$

- Oxidation of alcohols: Primary aliphatic and aromatic alcohols are oxidized to corresponding acids:

  Methanol → Formaldehyde → Formic acid

  Ethanol → Acetaldehyde → Acetic acid

  Benzoyal alcohol → Benzaldehyde → Benzoic acid

Methanol toxicity

Methanol has a relatively low toxicity compared to its products. Methanol is metabolized in the liver. In the first step of degradation, methanol is transformed to formaldehyde via the enzyme alcohol dehydrogenase (ADH). Transformation of formaldehyde to formic acid via the enzyme aldehyde dehydrogenase is faster. The metabolism of formic acid is very slow; thus, it often accumulates in the body, which results in metabolic acidosis. The major damage occurs to the optic nerve. Ethanol is given as an antidote, since it is the true substrate of alcohol dehydrogenase, methanol is spared.

Oxidation of aromatic hydrocarbons: Aromatic hydrocarbons are oxidized to phenolic compounds, which can further be conjugated with glucuronic aicd or sulfuric acid in phase 2 reactions so as to be excreted through urine.

Phase 1 reactions are mainly catalyzed by members of a class of enzymes referred to as mono-oxygenases, mixed function oxidases or cytochrome P450s. The reaction catalyzed by a mono-oxygenase (cytochrome P450) is as follows:

$RH + O_2 + NADPH + H^+ \rightarrow R\text{–}OH + H_2O +$ NADP

RH above can represent a very wide variety of xenobiotics, including drugs, carcinogens, pesticides, petroleum products, and pollutants.

(B) Reduction

Reduction does not occur extensively in human beings.

Examples: Reduction of Nitro compounds

- p-nitrobenzene → p-amino benzene
- p-nitrophenol → p-aminophenol

(C) Hydrolysis

Certain therapeutic compounds undergo hydrolysis.

Examples:

Acetyl salicylic acid → Acetic acid + Salicylic acid

Phase 2: Conjugation

Conjugation is a process by which the foreign molecules and their metabolites are coupled with a conjugating agent and are converted to soluble, non-toxic derivatives which are easily excreted in urine. Conjugation reaction **can occur independently or can follow phase 1 (hydroxylation) reactions**. Conjugation takes place primarily in liver but can occur in kidney also. After conjugation the products are generally rendered non-toxic but in certain conditions they are left unchanged or become more toxic.

Types of Phase 2 reaction:

1. Glucuronidation
2. Sulfation
3. Acetylation
4. Methylation
5. Conjugation with amino acids
6. Conjugation with G-SH

1. **Glucuronidation**: Glucuronidation is the most frequent conjugation reaction. **UDP-glucuronic acid**, is the glucuronyl donor, which is formed in the uronic acid pathway of glucose metabolism.

   Compounds conjugated with glucuronic acid are bilirubin, aromatic acids, benzoic acid, phenols, secondary and tertiary aliphatic alcohols, antibiotics like chloramphenicol, hormones—thyroid hormone, derivatives of corticosteroids and sex hormone metabolites, 2-acetylaminofluorene (a carcinogen), aniline, meprobamate (a tranquilizer).

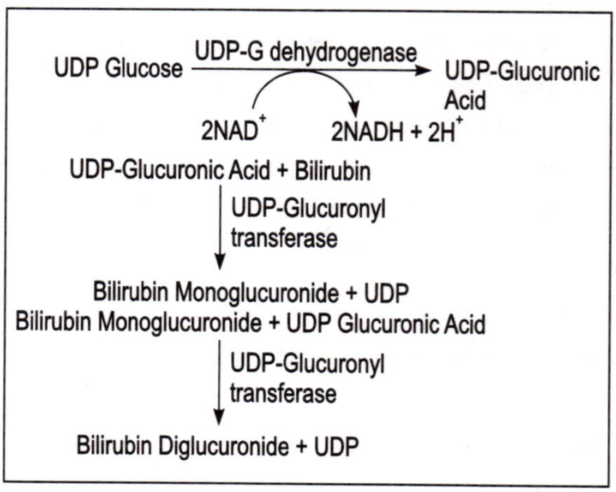

Fig. 29.1 : Bilirubin-UGT activity can be induced by a number of clincally useful drugs, including phenobarbital

2. **Sulfation**: The sulfate donor is adenosine 3′-phosphate-5′-phosphosulfate (PAPS). This compound is called "active sulfate". The enzyme is sulfo transferase. Compounds which are conjugated with sulphate are as follows: phenols, cresols, indole, steroids, oestrogen and androgens, tyrosine to form tyrosine-O-sulphate, which is required for the formation of fibrinogen, glycosaminoglycans, glycolipids, and glycoproteins.
3. **Acetylation**

- Acetylation is represented by

  X + Acetyl-CoA → Acetyl-X + CoA
- These reactions are catalyzed by **acetyltransferases** present in the cytosol of various tissues, particularly liver. Compounds conjugated by acetylation are sulphanilamide, PABA (Para Amino Benzoic Acid), isoniazid.

4. **Methylation**: Methylation is limited in the body. S-adenosyl methionine (active methionine) acts as a methyl group donor. Reactions are called Transmethylation reactions. Enzymes catalyzing the reactions are methyl transferases. Compounds conjugated by methylation are nicotinamide, O-methylation of estrogen, norepinephrine, epinephrine and their metabolites.
5. **Conjugation with amino acids**

(a) **Conjugation with glycine**

- Benzoic acid + glycine → Hippuric acid
- Cholic and deoxy cholic acid are conjugated to form glyco cholic acid and glycodeoxy cholic acid.

(b) **Conjugation with cysteine**: A few aromatic compounds are conjugated with cysteine in the presence of acetic acid to for mercapturic acid.

- Bromobenzene + Cysteine + Acetic acid → Bromophenyl mercapturic acid
- Naphthalene + Cysteine + Acetic acid → Nephthyl mercapturic acid

(c) **Conjugation with glutamine**

Phenyl acetic acid + Glutamine → Phenyl acetyl glutamine

This reaction is important in patients of phenyl ketonuria, since excess of phenyl acetyl glutamine is excreted in urine, that imparts a mousy odour to the urine.

6. **Conjugation with Glutathione**

- Glutathione (Y-glutamyl-cysteinylglycine) is a tripeptide consisting of glutamic acids, cysteine, and glycine. Glutathione is commonly abbreviated GSH (because of the sulfhydryl group of its cysteine, which is the business part of the molecule).
- A number of potentially toxic electrophilic xenobiotics (such as certain carcinogens) are conjugated to the nucleophilic GSH in reactions that can be represented as follows:

$$R + GSH \rightarrow R\text{-}S\text{-}G$$

where R = an electrophilic xenobiotic

INDEX

A

B

C

D

E

F

G

H

I

J

K

L

M

N

O

P

R

S

T

Some exciting gifts are up for grabs with this book!!

FREE Gifts Rs. 1000/-

Dear Readers

By now you must have gone through this book in the entirety. So we invite your assessment on this title with absolute impartiality and candid observations. After all, we publishers faithfully depend on your benchmarking. Friends, please fill up the open spaces below with CAPITAL LETTERS. You may add some extra sheets of paper if the space provided is not sufficient for you.

i. What are the important topics of your syllabus which are not adequately covered in this book?

ii. In which chapters and on what topics the treatment of the subject-matter by the author is not satisfactory and why?

iii. Have you come across the typos, factual inaccuracies, inconsistencies, and other forms of mistakes in the book? If yes, then please specify the mistakes and page numbers.

iv. Name top three books on the same subject in your consideration. Could you provide some reasons?

Your 1st Preference: ....

Your 2nd Preference: ....

Your 3rd Preference: ....

v. Any suggestion to improve the quality of illustrations, diagrams, etc.?

vi. How did you like the cover design? Any suggestion on colours?

vii. Suggestions for further improvement of production-standard:

viii. Who recommended you this book:

☐ Teacher ☐ Friend ☐ Parents ☐ Bookseller ☐ Librarian

ix. Name of the person who has recommended:

Designation: ....

Address: ....

Phone No.: ....

e-mail: ....

x. Name and address of the bookseller from whom you have purchased this book:

....

....

....

xi. Name and address of the library where you would like to find the book:

....

....

xii. Name and address of your institution (Please mention the University or Board):

....

....

xiii. Your name, e-mail, and complete postal address:

....

....

....

xiv. A few other titles published by us that you would like to read:

....

....

The best assessment will be rewarded half-yearly. The award will be in the form of our publications, as decided by the Editorial and Marketing Operations Division, amounting to Rs. 1000/- (Rupees one thousand) only.

....

Please mail the filled up coupon at your earliest to :

Editorial & Marketing Operations Division

NEW CENTRAL BOOK AGENCY (P) Ltd

8/1 Chintamoni Das Lane, Kolkata 700 009

West Bengal, INDIA

Phone + Fax: 033 + 2241 8411, 2241 8573, 2241 9094

e-mail: ncbapvtltd@eth.net, ncba@cal2.vsnl.net.in

Website: www.ncbaglobal.in